Machine Learning for Oracle Database Professionals

Deploying Model-Driven Applications and Automation Pipelines

Heli Helskyaho
Jean Yu
Kai Yu

Apress®

Machine Learning for Oracle Database Professionals: Deploying Model-Driven Applications and Automation Pipelines

Heli Helskyaho
Helsinki, Finland

Jean Yu
Austin, TX, USA

Kai Yu
Austin, TX, USA

ISBN-13 (pbk): 978-1-4842-7031-8
https://doi.org/10.1007/978-1-4842-7032-5

ISBN-13 (electronic): 978-1-4842-7032-5

Managing Director, Apress Media LLC: Welmoed Spahr
Acquisitions Editor: Jonathan Gennick
Development Editor: Laura Berendson
Coordinating Editor: Jill Balzano

Cover image designed by Freepik (www.freepik.com)

Distributed to the book trade worldwide by Springer Science+Business Media LLC, 1 New York Plaza, Suite 4600, New York, NY 10004. Phone 1-800-SPRINGER, fax (201) 348-4505, e-mail orders-ny@springer-sbm. com, or visit www.springeronline.com. Apress Media, LLC is a California LLC and the sole member (owner) is Springer Science + Business Media Finance Inc (SSBM Finance Inc). SSBM Finance Inc is a **Delaware** corporation.

For information on translations, please e-mail booktranslations@springernature.com; for reprint, paperback, or audio rights, please e-mail bookpermissions@springernature.com.

Apress titles may be purchased in bulk for academic, corporate, or promotional use. eBook versions and licenses are also available for most titles. For more information, reference our Print and eBook Bulk Sales web page at http://www.apress.com/bulk-sales.

Any source code or other supplementary material referenced by the author in this book is available to readers on GitHub via the book's product page, located at www.apress.com/9781484270318. For more detailed information, please visit http://www.apress.com/source-code.

Printed on acid-free paper

Table of Contents

About the Authors

Heli Helskyaho is the CEO of Miracle Finland Oy. She holds a master's degree in computer science from the University of Helsinki and specializes in databases. She is currently working on her doctoral studies, researching and teaching at the University of Helsinki. Her research areas cover big data, multi-model databases, schema discovery, and methods and tools for utilizing semi-structured data for decision making.

Heli has been working in IT since 1990. She has held several positions, but every role has included databases and database designing. She believes that understanding your data makes using the data much easier. She is an Oracle ACE Director, an Oracle Groundbreaker Ambassador, and a frequent speaker at many conferences. She is the author of several books and has been listed as one of the top 100 influencers in the IT sector in Finland for each year from 2015 to 2020.

Heli can be reached at `www.linkedin.com/in/helihelskyaho/`, `https://helifromfinland.blog`, and `https://twitter.com/HeliFromFinland`.

Jean Yu is a Senior Staff MLOps Engineer at Habana Labs, an Intel company. Prior to that, she was a Senior Data Engineer on the IBM Hybrid Cloud Management Data Science Team. Her current interests include deep learning, model productization, and distributed training of massive transformer-based language models. She holds a master's degree in computer science from the University of Texas at San Antonio. She has more than 25 years of experience in developing, deploying, and managing software applications, as well as in leading development teams. Her recent awards include an Outstanding Technical Achievement Award for significant innovation in Cloud Brokerage Cost and Asset Management products in 2019 as well as an Outstanding Technical Achievement Award for Innovation in the Delivery of Remote Maintenance Upgrade for Tivoli Storage Manager in 2011.

Jean is a master inventor with 14 patents granted. She has been a voting member of the IBM Invention Review Board from 2006 to 2020. She has been a speaker at conferences such as North Central Oracle Apps User Group Training Day 2019 and Collaborate 2020.

Jean can be reached at www.linkedin.com/in/jean-yu/.

Kai Yu is a Distinguished Engineer in the Dell Technical Leadership Community. He is responsible for providing technical leadership to Dell Oracle Solutions Engineering. He has more than 27 years of experience in architecting and implementing various IT solutions, specializing in Oracle databases, IT infrastructure, the cloud, business analytics, and machine learning.

Kai has been a frequent speaker at various IT/Oracle conferences, with over 200 presentations in more than 20 countries. He has authored 36 articles in technical journals such as *IEEE Transactions on Big Data* and co-authored the book *Expert Oracle RAC 12c* (Apress, 2013). He has been an Oracle ACE Director since 2010. He has served on the IOUG/Quest Conference committee, as the IOUG RAC SIG president, and as the IOUG CLOUG SIG co-founder and vice president. He received the 2011 OAUG Innovator of Year award and the 2012 Oracle Excellence Award for Technologist of the Year: Cloud Architect by *Oracle Magazine*. He holds master's degrees in computer science and engineering from the Huazhong University of Science and Technology and the University of Wyoming.

Kai can be reached at https://kyuoracleblog.wordpress.com, www.linkedin.com/in/kaiyu1, and https://twitter.com/ky_austin1.

About the Technical Reviewer

Adrian Png is a seasoned solutions architect with more than 20 years of experience working with clients to design and implement state-of-the-art infrastructure and applications. He earned a Master of Technology (Knowledge Engineering) degree from the National University of Singapore and has applied his knowledge and skills in artificial intelligence and machine learning in practice, notably in his paper "Primer design for Whole Genome Amplification using genetic algorithms" (*In Silico Biology*, 2006; 6(6): 505–14). Adrian is also trained and certified in several Oracle technologies, including Oracle Cloud Infrastructure, Oracle Autonomous Database, Oracle cloud-native services, Oracle Database, and Oracle Application Express. He is an Oracle ACE and is a recognized contributor to the Oracle community. Most recently, he co-authored the book *Getting Started with the Oracle Cloud Free Tier* (Apress, 2020), an indispensable reference for anyone just starting Oracle Cloud and wishing to get the most out of Oracle's cloud offerings.

Acknowledgments

I want to thank my family, Marko, Patrik, and Matias, for their continuous support during this project, and my parents for their encouragement throughout my life. You gave me the confidence to write this book.

Thank you, Jean and Kai, for writing this book with me! It was a great pleasure to work with you. You are great friends and extremely talented.

Special thank you to Charlie Berger and Adrian Png for the extremely valuable comments, guidance, and support during this project. We could not have been able to write this book without your help!

And thank you, Jonathan, Jill, and the rest of the Apress team!

—Heli Helskyaho

I want to thank my parents, my grandfather, and my middle school math teacher for encouraging me to pursue an engineering career. Special thank you to my husband Kai, and to my graduate advisors Dr. Steve Robbins and Dr. Kay Robbins of UTSA for introducing me to the software industry in 1995.

Thank you Heli and Kai for working with me on this amazing project. I greatly admire your dedication and passion for the Oracle Machine Learning community.

I'd like to extend a special thanks to our reviewer Adrian Png and Apress editors Jonathan Gennick and Jill Balzano. I appreciate your extremely valuable review comments. Your guidance and support made this project possible for a novice writer like myself. I learned so much. Thank you!

—Jean Yu

ACKNOWLEDGMENTS

I dedicate this book to the readers of this book. I want to thank all the people who have assisted with this book, especially the technical reviewer, Andrew Peng, Apress editors Jonathan Gennick and Jill Balzano, and the rest of the Apress team, for their great efforts and patience in transforming the technical content into a finished book.

I thank Heli and Jean for their great talents, dedication, and amazing teamwork. It has been my great honor to be part of this great team with them.

I also thank my wife, Jean, and my daughter, Jessica, for their continuous support during this project. I want to dedicate this book to my parents, who encouraged me to pursue my education and career in computer technology.

I thank my mentor, Dell Senior Fellow Jimmy Pike, and the Dell Technical Leadership Community, who have inspired me to pursue technical excellence and expand my expertise in AI and machine learning. I also want to thank my manager, Ibrahim Fashho, for his great inspiration and longtime support.

—Kai Yu

Introduction

This book helps database developers and DBAs gain a conceptual understanding of machine learning, including the methods, algorithms, the process, and deployment. The book covers Oracle Machine Learning (OML) technologies that enable machine learning with Oracle Database, including OML4SQL, OML Notebooks, OML4R, and OML4Py. *Machine Learning for Oracle Database Professionals* focuses on Oracle machine learning in Oracle autonomous databases, such as the Autonomous Data Warehouse (ADW) database as part of the ADW collaborative environment. This book also covers some advanced topics, such as delivery and automation pipelines in machine learning.

This book also provides practical implementation details through hands-on examples to show how to implement machine learning with OML with ADW and how to automate the deployment of machine learning. The primary goal is to bridge the gap between database development/management and machine learning by gaining practical knowledge of machine learning. As a seasoned database professional skilled in managing data, you can apply this knowledge by analyzing data in the same data management system. Through this book, three authors with rich experience in machine learning and database development and management take you on a journey from being a database developer or DBA to a data scientist.

Readers and Audiences

This book is written for

- Database developers and administrators who want to learn about machine learning

- Developers wanting to build models and applications using Oracle Database's built-in machine learning feature set

- Administrators tasked with supporting applications in Oracle Database and ADW that use the machine learning feature set

Readers will learn how to do the following.

- Build an automated pipeline that can detect and handle changes in data/model performance

- Develop and deploy machine learning projects in ADW

- Develop machine learning with Oracle Database using the built-in machine learning packages

- Analyze, develop, evaluate, and deploy various machine learning models using OML4R and OML4SQL

CHAPTER 1

Introduction to Machine Learning

We live in exciting times with smartphones and watches, smart clothes, robots, drones, face recognition, smart personal assistants, recommender systems, self-driving autonomous cars, and 24/7 service chatbots, all of which are *artificial intelligence* (AI). But what is intelligence? Intelligence might be defined as the ability to acquire and apply knowledge and skills, in other words, to learn and use the skills learned. Artificial intelligence is exactly that but done by computers and software. In real life, people would like to have intelligent machines that can do things people find boring, do inefficiently, or maybe cannot do at all. It could be an extension of human intelligence through using computers, which is artificial intelligence. The core of artificial intelligence is the ability to learn, acquire knowledge and skills, which is *machine learning*. In machine learning, the machine is learning, reasoning, and self-correcting. Arthur Samuel defined machine learning in 1959 as "a field of study that gives computers the ability to learn without being explicitly programmed," which defines machine learning very well.

Why Machine Learning?

When Arthur Samuel defined machine learning in 1959, a lot of the mathematics and statistics needed was already invented. Still, there was no technology nor enough data to get the theory to practice. Today, there are hardware solutions, including GPUs and TPUs for matrix calculation, inexpensive storage solutions for storing data, open data sets, pre-trained models for transfer learning, and so on. All this makes it possible to use machine learning in the most interesting and useful ways. But it is not only that we are now able to use machine learning; it is also necessary to use it. With its volume, velocity, variety,

© Heli Helskyaho, Jean Yu, Kai Yu 2021
H. Helskyaho et al., *Machine Learning for Oracle Database Professionals*,
https://doi.org/10.1007/978-1-4842-7032-5_1

veracity, viability, value, variability, and visualization, big data has made it necessary to change traditional data processing into something more efficient and faster: machine learning.

Machine learning is not a silver bullet, and it should not be seen as such. Machine learning should be used only when it brings value. Typical use cases are when the rules and equations are complex or constantly changing. If the rules are understandable and can be programmed with if-else-then structures, machine learning might not be the best solution.

Classic examples of machine learning use cases are image recognition, speech recognition, fraud detection, predicting shopping trends, spam filters, medical diagnosis, or robotics. Some examples of machine learning to businesses are churn prediction, predicting customer behavior, anticipating voluntary employee attrition, and cross and up-sell opportunities.

An important requirement for machine learning is that you have data; otherwise, it makes no sense. The data is given to the machine, or the machine produces it, as it does in reinforcement learning. The better the quality of the data is, the better it can be used by machine learning. But even though the data is of excellent quality and machine learning works like a charm, a machine learning prediction is never a fact; it is always a sophisticated guess. Sometimes that guess is good and even useful, but sometimes it is not.

Also, a well-working machine learning model will no longer work well if something has changed—perhaps there is more noise in the data, the amount of data is larger, or the quality of data has lessened. In other words, it is important to understand that machine learning models need to monitor their defined metrics to make sure they still work as planned and to tune them if necessary.

What Is Machine Learning?

Machine learning can be divided into different categories based on the nature of the training data, the problem type, and the technique used to solve it. This book divides machine learning into three main categories: supervised learning, unsupervised learning, and semi-supervised learning.

Supervised Learning

Supervised machine learning is supervised by a human. Typically, that means that somebody has labeled the data to show the output or the correct answer. For example, somebody manually checks 1000 pictures and labels them to identify which of the pictures show cats, dogs, or horses.

Supervised learning is used when there is enough high-quality data and you know the target (e.g., the data is labeled). The models are trained and tested on known input and known output data to predict future outputs on new data. When testing the models, the prediction is compared to the true output to evaluate the models. To make this process meaningful, the training data must separate from the data used for testing. Each model is built using a different algorithm. A model maps the data to the algorithm and produces the prediction. So, each algorithm is processing the data differently. Depending on the chosen metrics, the evaluation process defines which algorithm performed the best, and the model using that algorithm can be implemented into production. The selection of an algorithm depends on the data's size, the type of data, the insights you want to get from the data, or how those insights will be used. The decision is a trade-off between many things, such as the predictive accuracy on new data, the speed of training, memory usage, transparency (black box vs. "clear box," how decisions are made), or interpretability (the ability of a human to understand the model).

Regression and classification are the most common methods for supervised learning. Regression predicts numeric values and works with continuous data. Classification works with categorized data and classifies data points. So, if you want to predict a quantity, you should use regression. If you want to predict a class or a group, you should use classification. An example of regression is the price of a house over time. An example of classification is predicting a beer's evaluation by rating it against other beers on a scale of 1 to 5, with 1 being poor quality and 5 being excellent. Figure 1-1 is a simple example of regression. From the line shown in Figure 1-1, you can see that for value 3, the prediction of the target value is 1.5.

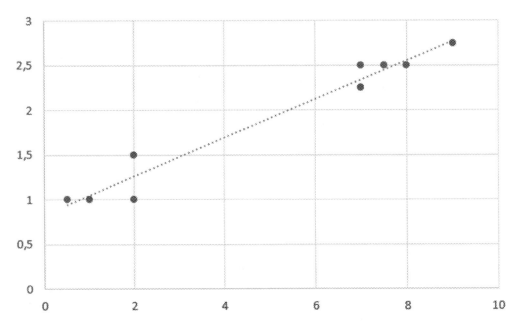

Figure 1-1. *An example of regression*

Figure 1-2 is an example of classification. The data points are classified in orange and blue. The red line shows in which category each data point belongs. You can see that point (4,1) belongs to the orange group, and point (9,2) belongs to the blue group.

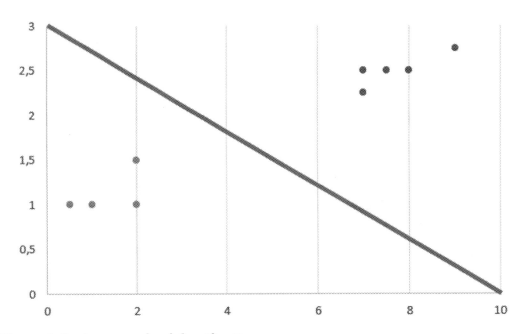

Figure 1-2. *An example of classification*

4

Time series forecasting can be a supervised learning problem. The machine learning model predicts the value of the next time step by using the value of a previous time step. You need data that is suitable for the purpose. This method is called the *sliding window method*. For example, the following is a small part of a data set.

Time	Measure
1	100
2	150
3	170
4	250
5	330

You can reconstruct this data set to be useful in supervised learning by setting the next value as the prediction of the value, as follows.

X	Y
?	100
100	150
150	170
170	250
250	330
330	?

The first and the last rows cannot be used because some of the information is missing, so we remove those rows. Afterward, there is a solid data set that can be used in supervised machine learning.

Time series forecasting can be used in weather forecasting, inventory planning, or resource allocation, for example. Time series prediction can be very complex, and understanding the data is very important. For example, trends in data might be different in summer than in winter, or on weekdays than on weekends. That must be considered when building the model or maybe several models for different trends.

Deep learning has become very popular as a technique for mainly supervised machine learning. Deep learning is typically used with more complex machine learning tasks on text, voice, recommender systems, or images and videos. Text can be transformed into speech using deep learning. Speech can be transformed into text, which can be used as input to another machine learning task, such as translating from the Finnish language to English.

Automatic speech recognition or natural language processing might also be tasks for deep learning. Recommender systems are producing recommendations for users to make their decision process easier and more fluent. There are three kinds of recommender systems: collaborative filtering, content-based, and hybrid recommender systems. A collaborative filtering recommender system uses the decisions of other users with a similar profile as a base for a recommendation for another user. Content-based recommender systems create recommendations based on similarities of new items to those that the user liked in the past. Hybrid recommender systems use multiple approaches when creating recommendations. Visual recognition and computer vision are very typical and useful tasks for deep learning. Image or action classification, object detection or recognition, image captioning, or image segmentation are useful in machine learning.

One difference between classical supervised learning and deep learning is that in deep learning you do not need to perform feature extraction at all, it is done by the machine as part of the deep learning process. In supervised learning, feature extracting is time-consuming manual work. Of course, that means that deep learning needs more data to do it and, in general, more resources and time. Deep learning has become more popular and useful because of so many improvements in different areas. There is a lot of digital data (photos, videos, voice, etc.) available. The technology has improved: existing data sets and pre-trained models, transfer learning, research such as combining convolutional layers to a neural network, and much more is available. Things that were difficult or nearly impossible to perform using deep learning have become easy and almost trivial. There are plenty of example codes that programmers can use and start building their first deep learning projects.

Deep learning uses neural networks for the prediction process and backpropagation to learn (e.g., tune the network). A neural network consists of neurons. Each input is multiplied by its weight, and a bias is added to that. When using an activation function, an output is passed to the next layer until the last layer and the prediction are reached. The weight and the bias are called *hyperparameters*. Their values are defined before the machine learning process starts. The first values are a guess, but by using backpropagation and an optimizer function, the process tunes those hyperparameters to have a better-performing model.

In a neural network, there are plenty of hyperparameters that need to be defined before starting the process, and they need to be tuned during it. Some examples of hyperparameters are the number of layers, number of epochs, the batch size, number of neurons in each layer, or what activation function, optimizer, and loss function to use. The backpropagation computes the loss function for the initial guess and the gradient of the loss function. Using that information the optimizer takes the steps to a negative gradient direction to reduce loss. This is done as long as needed to get the weights as good as possible. A convolutional neural network complements the neural network with convolutional layers. Convolutional neural networks are especially useful with image processing. A convolutional neural network consists of several convolutional layers (filter, output, pooling) and a flattening layer to pass the data to a neural network for further processing.

Algorithms for Supervised Learning

A model uses an algorithm to produce a prediction. The goal is to find the best algorithm for the use case. There are plenty of algorithms to be used with supervised learning.

For classification, examples of algorithms include k-nearest neighbors (kNN), naïve Bayes, neural networks, decision trees, or support-vector machine (SVM). kNN categorizes objects based on the classes of their nearest neighbors that have already been categorized. It assumes that objects near each other are similar. kNN is a simple algorithm, but it consumes a lot of memory, and the prediction speed can be slow if the amount of data is large or several dimensions are used. Naïve Bayes assumes that the presence (or absence) of a particular feature of a class is unrelated to the presence (or absence) of any other feature when the class is defined. It classifies new data based on the highest probability of its belonging to a particular class. For example, if a fruit is red, it could be an apple, and if a fruit is round, it could be an apple, but if it is both red and round, there is a stronger probability that the fruit is an apple.

Naïve Bayes works well for a data set containing many features (e.g., the dimensionality of the inputs is high). It is simple to implement and easy to interpret. A neural network imitates the way biological nervous systems and the brain process information. A large number of highly interconnected processing elements (neurons) work together to solve specific problems. Neural networks are good for modeling highly nonlinear systems when the interpretability of the model is not important. They are useful when data is available incrementally, you wish to constantly update the model, and unexpected changes in your input data may occur.

Decision trees are very typical classification algorithms. Decision trees, bagged decision trees, or boosted decision trees are tree structures that consist of branching conditions. They predict responses to data by following the decisions in the tree from the root down to a leaf node.

A *bagged decision tree* consists of several trees that are trained independently on data. Boosting involves reweighting misclassified events and building a new tree with reweighted events. Decision trees are used when there is a need for an algorithm that is easy to interpret and fast to fit, and you want to minimize memory usage but high predictive accuracy is not a requirement and the time taken to train a model is less of a concern. A *support-vector machine* (SVM) classifies data by finding the linear decision boundary, or hyperplane, that separates all the data points of one class from those of another class. The best hyperplane for an SVM is the one with the largest margin between the two classes when the data is linearly separable. If the data is not linearly separable, a loss function penalizes points on the wrong side of the hyperplane. Sometimes SVMs use a kernel to transform nonlinearly separable data into higher dimensions where a linear decision boundary can be found. SVMs work the best for high-dimensional, nonlinearly separable data that has exactly two classes. For multiclass classification, it can be used with a technique called *error-correcting output codes*. It is very useful as a simple classifier, it is easy to interpret, and it is accurate.

For regression tasks, some examples of algorithms are linear regression, nonlinear regression, generalized linear model (GLM), Gaussian process regression (GPR), regression tree, or support-vector regression (SVR).

Linear regression describes a continuous response variable as a linear function of one or more predictor variables. Linear regression could be used when you need an algorithm that is easy to interpret and fast to fit. It is often the first model to be fitted to a new data set and could be used as a baseline for evaluating other, more complex, regression models.

Nonlinear regression describes nonlinear relationships in data. It can be used when data has nonlinear trends and cannot be easily transformed into a linear space.

GLM is a special nonlinear model that uses linear methods. It fits a linear combination of the input to a nonlinear function of the output. It could be used when the response variables have non-normal distributions.

GPR is for nonparametric models used to predict the value of a continuous response variable; for example, to interpolate spatial data, as a surrogate model to optimize complex designs such as automotive engines, or to forecast mortality rates.

Regression trees are similar to decision trees for classification, but they are modified to predict continuous responses. They could be used when predictors are categorical (discrete) or behave nonlinearly.

SVM regression algorithms (SVR) work like SVM classification algorithms but are modified to predict a continuous response. Instead of finding a hyperplane that separates data, SVR algorithms find the decision boundaries and data points inside those boundaries. SVR can be useful with high-dimensional data.

Unsupervised Learning

Unsupervised learning is machine learning with unlabeled data, with an unknown target, to find something useful from the data. Unsupervised learning finds hidden patterns or intrinsic structures in input data.

Clustering is one of the most common methods for unsupervised learning. It is used for exploratory data analysis to find hidden patterns or groupings in data. There are typically two ways of clustering: hard and soft. In hard clustering, each data point belongs to only one cluster, whereas in soft clustering, each data point can belong to more than one cluster.

In Figure 1-3, you can see data points, and in Figure 1-4, you see how they have been clustered in two clusters: green and blue. The idea of clustering is that you tell the algorithm that you want to break the data into two groups, and it finds things that are common to the data points and things that are different. Using that information, the algorithm decides which group (cluster) a particular data point belongs to.

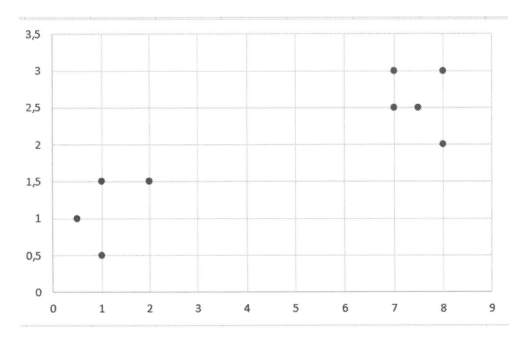

Figure 1-3. *Data points for clustering*

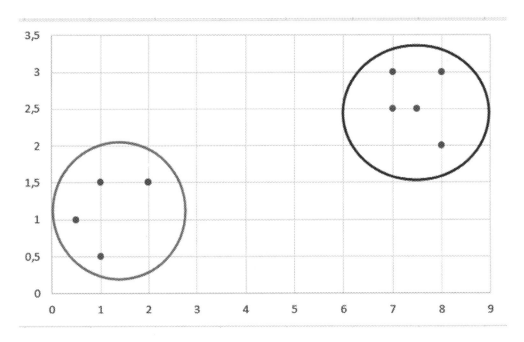

Figure 1-4. *Data points clustered in two clusters: green and blue*

Association rule mining/learning is also a very useful unsupervised machine learning method. It is for identifying hidden patterns in the data set. A typical example of association rules is shopping basket data analysis. Using association rules, you can find interesting patterns on what people typically buy. Using that information, you can better plan the layout of a shop, product placement in a supermarket, or launch better market campaigns to different customer groups. A classic example is that diapers and beer are often bought at the same time. The historical explanation was that fathers stop by the store to buy diapers on their way home from the office and often end up buying beer as well.

Anomaly detection is for finding anomalies in a data set. It can be used with unsupervised, supervised, or semi-supervised machine learning. Figure 1-5 is an example of an anomaly (marked in red). Anomaly detection can be used for finding exceptions that must be handled, or it can be used for data preparation for machine learning to find outliers. For instance, if an anomaly has been detected in a log file, it can alert the database administrator to check what is going on.

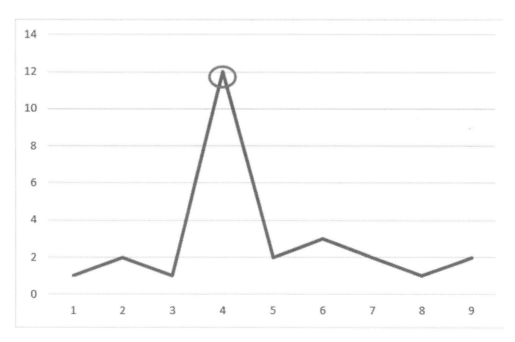

Figure 1-5. *An example of an anomaly*

Dimensionality reduction is an unsupervised machine learning method that can be used when there is a need to reduce the data set's features as preprocessing for supervised learning or to tune the model. Feature selection is a technique to reduce the number of features in a data set to improve the model accuracy or performance.

It ranks the importance of the existing features in the data set and discards less important ones. Feature extraction also aims to reduce the number of features in a data set, but it creates new features from the existing ones and then discards the original features. A term used for both methods is *dimensionality reduction* because they aim to reduce the dimensions/features of the data set.

Algorithms for Unsupervised Learning

There are plenty of algorithms for unsupervised learning. For hard clustering, several algorithms are available, including k-means, k-medoids, and hierarchical clustering. K-means (Lloyd's algorithm) partitions data into k number of mutually exclusive clusters (centroids) and assigns each observation to the closest cluster. Then it moves the centroids to the true mean of its observations. *K-means* works well when the number of clusters is known, and there is a need for fast clustering of large data sets.

K-medoids is similar to k-means but requires that the cluster centers coincide with points in the data. In other words, it chooses datapoints as centers, medoids. k-medoids can be more robust to noise and outliers than k-means. It works well when the number of clusters is known, and there is a need for fast clustering of categorical data.

Hierarchical clustering is used when the number of clusters is unknown and/or you want visualization to guide the selection. It works either as a divisive or agglomerative method. A divisive method assigns all observations to one cluster and then partitioning that cluster into several clusters. An agglomerative method sets each observation to its own cluster and merges similar clusters. Of these two types of methods, the agglomerative method is used more often.

For soft clustering, there are algorithms like Fuzzy C-means (FCM) or Gaussian mixture model. FCM is similar to k-means, but data points may belong to more than one cluster. It could be used when the number of clusters is known and they overlap. A typical use case is pattern recognition. A Gaussian mixture model is a partition-based clustering where data points come from different multivariate normal distributions with certain probabilities, such as home prices in different areas. It could be used when data points might belong to more than one cluster, and those clusters have different sizes and correlation structures within them.

For anomaly detection, kNN is a simple algorithm, and for each data point, its distance to its k-nearest neighbor could be viewed as the outlier score.

An Apriori algorithm is for association rule-learning problems. It identifies items that often occur together.

Semi-Supervised Learning

There is also a combination of unsupervised and supervised machine learning called *semi-supervised learning*. In semi-supervised learning, a small part of the data is supervised data with labeled data, and a large part of the data is unsupervised data with unlabeled data. The idea in semi-supervised learning is that you should find a way to let the algorithm label the data automatically. These algorithms are often based on assumptions on the relationships of the data distribution. Some algorithms assume that the points close to each other are likely to have similar or the same labels (continuity assumption). Some assume that data in the same cluster most likely have the same labels (cluster assumption). For high-dimensional data, manifold assumption might help find the label.

There are many algorithms and techniques to automatically label the data, and more techniques will soon be invented.

A simple way to label the unlabeled data is called *pseudo-labeling*. The process of pseudo-labeling is simple.

1. Train the model with a small set of labeled data.

2. Use the model to label unlabeled data. Set the prediction as the label. This is called a *pseudo-label*.

3. Link the labels from the labeled training data with the pseudo-labeled data. Link the input data from the labeled data with the unlabeled data.

4. Train the model with that data set.

After this process, the accuracy of the model should be better than simply training it with the labeled data set.

Active learning is a technique in which the learning algorithm chooses a subset of unlabeled data and queries a user interactively to label it. And since the algorithms chose the data set to be labeled, it is assumed this data set is very useful in learning the algorithm.

Many interesting and promising areas in machine learning research have something to do with semi-supervised learning. Examples of those areas are reinforcement learning and self-supervised learning (self-supervision). Some people include reinforcement learning into semi-supervised learning. Some say it is a different concept because it is more about learning a skill than labeling data and should have its own category. This book includes it in semi-supervised learning to keep the concepts simple.

Reinforcement Learning

Reinforcement learning (RL) has a different approach to machine learning. Its goal is to get the maximum reward, not to predict the input data's output. RL is typically used in use cases where the computer needs to learn a skill, such as playing chess or parking a car. RL works well with problems like games, optimization, or navigation because data can be generated easily. A computer can play against itself or a human player and generate a lot of data about the game's strategies. It learns while playing, and it can also use transfer learning to learn faster. RL's best feature is that it can become better than a human since humans are not teaching it with their limited skills. It takes hundreds of thousands of mistakes to learn, which is not possible in all use cases.

A *Markov decision process* (MDP) is the basis of reinforcement learning (RL) and the mathematical formulation for an RL problem. MDP provides a mathematical framework (a discrete-time stochastic control process) for modeling decision making when outcomes are partly random and partly under the control of a decision-maker, just like in RL. The RL process is shown in Figure 1-6.

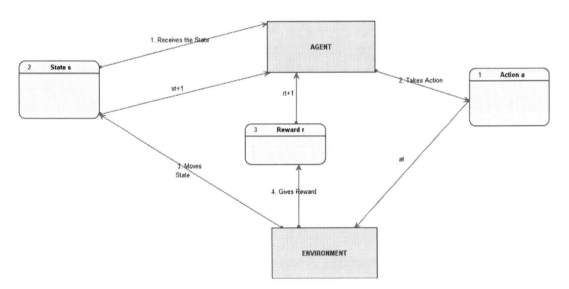

Figure 1-6. *The reinforcement learning process*

A decision-maker/learner is defined as an agent, and anything outside the agent is defined as an environment. The interactions between an agent and an environment are described by three core elements: state (s), action (a), and reward (r). The agent receives as input the current state of the environment. Then the agent chooses an action.

The action changes the state of the environment. The value of this state transition is communicated to the agent through a scalar reinforcement signal, the reward. The state of the environment at a moment in time (t) is denoted as s_t. The agent examines s_t and performs a corresponding action a_t. The environment then alters its state s_t to s_{t+1} and provides a feedback signal r_{t+1} to the agent. The agent receives a feedback signal for every time-step t until it reaches the terminal state s_T. The agent learns using these reward signals, and its behavior (B) aims to as large a total sum of reward signal values as possible. The better the signals have been planned, the faster the agent learns. These signals are defined as reward functions. The agent performs the task many times to build a policy network, including the possible actions it could take. The network is tuned using policy gradients and probabilities. The main goal is to find the optimal policy for an agent: a function that defines the actions the agent should take in each state to gain the maximum sum of rewards.

There are several problems in defining such a policy. The first problem is to define a reward function that teaches the agent the correct behavior. If the reward is given only at the end of the task, the agent doesn't know what action caused the reward. This is called a *sparse reward*. The problem is called a *delayed reward* or a *credit assignment problem*. Also, the algorithm is inefficient, and learning takes a very long time. It might be that the agent can never finish the task because it is too complicated, and therefore it never gets a reward and never learns. If the reward is given after each action and the reward function is not defined correctly, it might be teaching the agent a wrong behavior. It also limits learning to the human level, which is not what we want. Another problem related to this solution is *reward shaping*, since having the reward after each action means it is a custom process for a certain environment, it is not scalable. Defining the policy can be difficult because you do not know if there will be an infinite horizon (stationary) or a limited number of steps. Knowing the maximum number of actions leads to different policies, set of actions, and utility of sequences. It changes MDP to a Markov chain.

Multi-agent reinforcement learning (MARL) is a reinforcement system that consists of more than one agent. Depending on the task, those agents can be independent, cooperate, or compete. They can either share what they have learned with the other agents or choose not to share it. MARL solves complex tasks. Two advantages are that it is scalable and reusable. The challenges include defining and achieving the global goals from local actions, learning while the environment is constantly changing, and finding the best way and timing for reward assignment—a hierarchical reinforcement learning (HRL) approach to solving long-horizon problems with sparse and delayed rewards. With HRL, subtasks and abstract actions can be used in different tasks in the same domain (transfer learning).

Self-Supervised Learning

The next big step in machine learning, creating machines that can learn (and think) like a human, is self-supervised learning. In self-supervised learning, the machine learns by predicting anything missing from all the information it has: future from past, past from present, or top from bottom. In self-supervised learning, the learning model trains itself by leveraging one part of the data to predict another part. The idea is that it generates labels accurately by itself without human labeling data. In other words, it converts unsupervised learning, without labels, to supervised learning, with labels. A picture or a text might be missing a part, and self-supervised learning is used to automatically fill in the missing part.

The Machine Learning Process

The first step in a machine learning process is to define the problem or the task you are trying to solve. Then you find data that has the answer to that task. The machine learning technique and the problem type are determined based on the task. The technique can be unsupervised, supervised, or semi-supervised. The problem type, called a *method* or a *function*, could be, for example, classification, regression, anomaly detection, or clustering. The problem type and chosen technology define the possible algorithms available to be used for solving such problem type.

Hyperparameters are another set of parameters that depend on the method. They are parameters whose values are set prior to the commencement of the learning process. A typical hyperparameter for unsupervised learning and clustering is the number of clusters. Models are built using a carefully selected set of features, tuned hyperparameters, and available algorithms for the selected problem type. These models are compared and evaluated, and then the best model is deployed to production. Of course, none of them may be good enough. The reason for that could be that the task was not defined correctly, or there was something wrong with the data—it was not good enough, there was not enough of it, or it did not include the answer to the business question. Or perhaps the business problem cannot be tackled using machine learning.

If a task is to find something useful from data, it could be a use case for unsupervised machine learning. Defining a task can be difficult since it needs to be precise. This is made more difficult because it is defined using a human language, which is not often precise.

The task and the training data available guide the decision on what method to use. The first question is whether the training data is labeled with the outcome or not? If it is labeled, you could use supervised learning. If not, you should choose unsupervised learning. If the data is partially labeled, semi-supervised learning could be the option. The next question is the type of outcome, the prediction. Are the outcomes discrete or continuous? If you choose supervised learning, your option for discrete outcome could be classification and for continuous data regression. For unsupervised learning clustering would be an option for discrete data and dimensionality reduction for continuous data. If the data is partially labeled and you chose semi-supervised learning, the question is what kind of technique to use to let the machine label the data by itself. This is one of the most interesting and developing areas in machine learning.

The data consists of rows representing the observations (cases) and the features, such as the size of a house or the age of a person. Features are sometimes called *attributes* or *columns* because they are the columns in a table. There are slight semantic differences between these terms but is general language they are used as synonyms. If you want to perform predictions, the most important feature is the target—the feature whose value you are trying to predict. Feature selection finds the features that are meaningful for the task at hand. The fewer features are needed, the more efficient the learning process is. However, if there are not enough features, the learning might result in non-accurate prediction.

You should carefully check the data's quality while preparing the data for machine learning algorithms. The quality of the data defines the machine learning process's success. The first step in preparing the data is feature engineering. It turns data into information that a machine learning algorithm can use. This step includes feature selection, feature generation, feature extraction, and feature transformation, among others.

The process starts with understanding the features and the interaction between them. There are three kinds of features: data, model, and target. We try to predict a target feature's value, such as the temperature on Monday or whether an employee is planning to leave the company. Data features are those columns in the data set that build, test, or score a model. In other words, not all the columns in a table or view need or should be used in building the model, only those that affect the prediction.

Feature selection identifies the most relevant features that provide the best predictive model for the data. This step is important because if the model has too few features, it will not predict well enough or incorrectly. And if it has too many features, it will be too slow. You want to know that a feature is relevant and independent from other features; and if it not, which of the other features should be selected. For example, if trying to predict a house's sale price, the house's size and the number of rooms are definitely features somehow dependent on each other. The desion to make is should you use one of them (which one?), both of them, or discard both and create a new feature based on the information they contain.

The knowledge about feature correlation can be used to check the data quality for instance, a house with five rooms cannot be only 30 square meters in size. A correlation plot is a useful tool to see how different features correlate. For each feature you should also understand if that feature is simple or not, and can it be compared to other instances of the same feature. For instance, a house's GPS coordinates can be compared to each other's, but it usually makes no sense. In this case, perhaps the GPS coordinates can be used for something useful that will affect the price of the house; for example, the distance to the beach or the city's center. Conversion often includes converting the datatype, such as age to child/adult or date of birth to age. Sometimes the data needs to be modified because it is unnecessarily detailed. You might want to remove details by discretization; for example, the actual weight of an element is 11.1 kg, but 10 kg is enough information.

It is important that the datatype of a feature is correct for the algorithm. Numeric implies that the data can be put in order based on its value. If the datatype is defined as numeric, but the data is categorical, it can cause confusion. For example, a zip code could be defined as numeric, but it is not something that you should do because the order of zip codes is meaningless. Instead, a zip code should be categorical data. The model will find better use if it is converted to char datatype.

Usually, dates and times are very complex to handle since the way they are stored in a data source might vary. In a database, this problem is usually smaller since it is designed to use the correct datatype. A database also has a lot of functions for converting date/time data. You could transform the date to the name of the weekday or define the duration of an event based on the start and end dates.

When processing a text datatype, the text is tokenized; in other words, it is divided into parts (tokens). A token might be a word or a combination of words. Tokenizing is very complex and heavily involved with the language used. After the text is tokenized, it is normalized.

If the data includes transactional data, you might want to de-normalize it. In transactional data, the information for each case consists of multiple rows. For example, your phone call data is associated with your phone number, but there is transactional data about all the calls you have made. All this data has to be brought to the case level (your phone) by de-normalizing the call data. That would mean creating nested columns.

Finding outliers is important since outliers can have a skewing effect on the data. A value is considered as an outlier if it deviates significantly from most other values of the feature. Usually, a linear plot is a very useful tool to find outliers. Deciding what to do with outliers depends on whether the outlier is seen as problematic or perfectly valid data: keep, delete, or change. Outliers can interfere with the effectiveness of transformations such as normalization or binning. Normalization is a technique for reducing the range of numerical data. Binning (discretization) is a technique for reducing the cardinality of continuous and discrete data. These techniques can improve model quality by strengthening the relationship between attributes.

Maybe the biggest problem with data is missing data. When data is missing, you can either ignore and discard data or do imputations. Imputation means that the missing values in a data set are replaced with plausible values. If you choose to ignore and discard data, you can either do it for all the missing data in a data set or to some instances or features. But before you decide to ignore or discard any data, it is important to do a feature importance analysis to understand the relevance of that feature to the machine learning model. If the feature is not relevant to the prediction, you can easily discard it; otherwise, it is important to find a valid value for the missing data.

The goal is to employ known relationships that can be identified in the valid values of the dataset to assist in estimating the missing values. Imputation aims to fill in the missing values with estimated ones. The solution for imputation can be found by investigating the randomness of missing data. Missing data is random when the missing value for a feature does not depend on either the known values or the missing data. In this case, any missing data treatment method can be applied without the risk of introducing bias on the data. In many cases, features are not independent of each other and, through the identification of relationships among features, missing values can be determined. But the question is what data should be imputed. It might be the most probable value of that feature, a mean of the existing values, the value of the previous/next feature, or something else.

To make that decision, you must understand the data well. Mean and mode imputation are frequently used methods for imputation. It replaces the missing data for a given quantitative feature by the mean and a qualitative feature by the mode of all known values of that feature. A missing value can be estimated using a hot or a cold deck or a prediction model. In the hot deck, the missing value is estimated from the current data. In the cold deck, it's done using another data set. In both methods, the data is partitioned into clusters. With clustering, each instance with missing data is associated with one cluster. The complete cases in that cluster fill in the missing values using the mean or mode of the feature within a cluster.

Prediction models are the most sophisticated way to handle missing data. You create a predictive model to estimate values to substitute the missing data. This only works if the features have correlations among themselves.

When a task is described and understood, and the data is collected and prepared, it's time to build the models using available algorithms for the method. The algorithms available depends on the technology and programming language. When the models have been built, you need to evaluate the models to see if they are good enough for the purpose.

The evaluation depends on the chosen metrics. The metrics could be accuracy or error rate, for example. Several tools can be used for evaluation, including confusion matrix, ROC curve, or coverage curve. We talk more about evaluation later in this book.

If the model is not good enough, it might be improved using feature engineering or hyperparameter tuning. Feature engineering can use feature selection techniques or feature transformation/extraction techniques. There are techniques like stepwise regression, sequential feature selection, regularization, or neighborhood component analysis for feature selection.

Stepwise regression adds or removes features/attributes sequentially until there is no improvement in prediction accuracy. *Sequential feature* selection adds or removes predictor features iteratively and evaluates the effect of each change on the performance of the model. *Regularization* uses shrinkage estimators to remove redundant features by reducing their weights (coefficients) to zero. *Neighborhood component analysis* (NCA) finds the weight each feature has in predicting the output so that features with lower weights can be discarded.

Feature transformation is a form of dimensionality reduction that is used when data sets get bigger and there is a need to reduce the number of features or dimensionality. This technique includes principal component analysis (PCA), factor analysis, and non-negative matrix factorization (NNMF).

PCA converts a set of observations of possibly correlated variables into a smaller set of values of linearly uncorrelated variables called *principal components*. The first principal component captures the most variance, followed by the second principal component, and so on. Factor analysis identifies underlying correlations between features in a data set to provide a representation in terms of a smaller number of unobserved features, factors.

NNMF is also known as *non-negative matrix approximation*. It is used when model elements must represent non-negative quantities, such as physical quantities.

Hyperparameter tuning, also called *hyperparameter optimization*, chooses an optimal set of hyperparameters for a learning algorithm. Hyperparameters control how a machine learning algorithm fits the model to the data. Tuning hyperparameters is an iterative process. It starts by setting parameters based on the best guess and finding the best possible values to yield the best model. As you adjust hyperparameters and the model's performance begins to improve, you see which settings are effective and still require tuning. Grid search, Bayesian optimization, gradient-based optimization, and random search are examples of optimization algorithms. In general, it is much better to tune the parameters than to find more complex algorithms. A simple algorithm with well-tuned parameters is often better than an inadequately tuned complex algorithm in many ways.

After a while, the well-performing model is not as good as it was and should be. Maybe the data has more noise than it had when the model was trained, maybe the data has somehow changed, maybe there is just more data than before, or it could be anything that has changed the model performance or other measurement factor. Then there is a need for model improvement again. Model improvement cannot be made without measuring. If you do not have measures to follow, you do not know when there is a need to improve or whether the actions improve or worsen the metric.

A very typical measurement is accuracy. One commonly used technique for testing model accuracy is *cross-validation*, sometimes called *rotation estimation*. In cross-validation, the data is divided into *n* parts of equal size. Each part is used for training and testing. For instance, if we divide the data into ten parts, parts 1–9 are used for training, and part 10 is used for testing. Then we rotate, and parts 2–10 are used for training, and part 1 is used for testing. We rotate again, and parts 3–10 and 1 are used for training, and part 2 is used for testing. We keep rotating until each part has been in each position.

Many machine learning tools have some kind of AutoML functionalities that use machine learning to choose and optimize the machine-learning pipeline. That technique is called *meta-learning*. Depending on the tool, it might automate feature engineering, feature analysis and evaluation, transformations, model evaluations, and so on.

Machine learning operations (MLOps) enables data science and IT teams to collaborate and adds discipline to the model development and deployment by defining processes, including monitoring, validation, and governance of machine learning models. It is also important to note that machine learning is all about data. Often, data is sensitive, or combining different data makes non-sensitive data very sensitive. Security is a very big part of machine learning and should be included in the process from beginning to end.

Another aspect of data usage in machine learning is ethics. Data can be used for good or for bad. That decision must be made every time a new model is used. Only humans can make the decisions. For example, a surveillance camera has data on faces recognized by a machine learning model. This data can be used for good (e.g., finding a lost child) or for bad (e.g., controlling people). Another problem with data is that it can be biased. If a model is trained with biased data, the model learns to be biased too. This should be recognized, and the data should be fixed before deploying the model.

Summary

Machine learning is about data, features, algorithms, and models. The process starts with defining the task, the business problem you want to solve with machine learning, and finding the data. Finding data that supports the task might be difficult, and finding enough high-quality data might be even more difficult. When data has been identified, it usually needs to be cleaned and wrangled into a form that is usable by the algorithms. This phase might include processing null values, transforming, consolidating, augmenting, standardizing, filtering, or in any other way transforming the data to a form that is the most useful for the purpose. This step usually takes a lot of time. This is one reason many machine learning tool providers offer tools that can automate part of or almost the whole machine learning process.

Machine learning problems can be divided into types based on the data and the problem to be solved. A model addresses the task. But the model is only as good as its features. That's why feature engineering is an important part of the process before training, testing, and implementing the model. The model maps the data to an algorithm. There are many possible algorithms for all possible machine learning problem types that would be easy to make a process complicated. Therefore, it is good to keep in mind that a simple algorithm with well-tuned parameters often produces a better model than an inadequately tuned complex algorithm.

CHAPTER 2

Oracle and Machine Learning

Machine learning is a broad area, and there are different users for it. To support people with different backgrounds and preferences, Oracle offers several tools for machine learning. Most of these tools have been implemented to the Oracle Database kernel, allowing efficient and safe data processing. Oracle Database also makes it possible to use all its features and all the data models in a converged database, such as JSON, graph, and spatial. You can mix and match these technologies to build the toolset for machine learning that best suits your needs. This chapter introduces some of the most useful technologies Oracle offers for machine learning.

Oracle Machine Learning for SQL (OML4SQL)

Oracle Machine Learning for SQL (OML4SQL) is a very good option for an Oracle professional familiar with SQL and PL/SQL. The power of OML4SQL is that there is no need to move the data anywhere since the machine learning algorithms are in the database core and use the data there. The algorithms are limited, but in most cases, they are enough. You can process structured data and unstructured text data using OML4SQL. If you need to build models on images and videos, OML4SQL is not the right tool. And, OML4SQL does not support reinforcement learning. Still, it is good for the most common unsupervised and supervised machine learning problems, including deep learning with textual data.

OML4SQL comes with the Oracle Database and is free to use. It consists of three PL/SQL packages: DBMS_PREDICTIVE_ANALYTICS, DBMS_DATA_MINING_TRANSFORMING, and DBMS_DATA_MINING. OML4SQL can be used in Oracle SQL Developer, Oracle SQL Developer Web, Oracle Machine Learning Notebooks in Oracle

© Heli Helskyaho, Jean Yu, Kai Yu 2021
H. Helskyaho et al., *Machine Learning for Oracle Database Professionals*,
https://doi.org/10.1007/978-1-4842-7032-5_2

Autonomous Database, or any interface that lets you call PL/SQL packages. Oracle SQL Developer also has an add-on for an OML4SQL graphical user interface. We talk more about OML4SQL in Chapter 3.

Oracle and Other Programming Languages for Machine Learning

Most programming languages have drivers/libraries for using data from an Oracle Database. This section discusses R, Python, and Java.

R

R has been available in Oracle Database for a long time. Using special Oracle libraries, you can use Oracle tables as data frames in R. You can embed R in SQL and PL/SQL to build models or batch score data sets, and much more. Oracle has a special version of R, Oracle R Distribution (ORD), called Oracle R Enterprise (ORE). It can be used in the database with Oracle Machine Learning for R (OML4R) or outside the database with the R API.

Oracle Machine Learning for Spark (OML4Spark) is for performing R using data storage technologies called *data lakes*. In short, it combines the advantages of R with the power and scalability of Oracle Database and data lake environments (Apache Hive, Apache Impala, Apache Spark, or Hadoop). OML4R is included with the Oracle Database. OML4Spark is a component of the Oracle Big Data Connectors software suite for on-premise big data solutions, and it is included with Oracle Big Data Service.

ROracle is an R package enabling scalable and performant connectivity to an Oracle database. It is found in CRAN (Comprehensive R Archive Network) and can be used with any R IDE. Two other packages, RODBC and RJDBC, can connect to an Oracle Database, but they do not offer any of the benefits that ROracle offers in performance and scalability.

Oracle Machine Learning for R (OML4R) is R inside an Oracle Database. You can use it to build parallel and distributed machine learning algorithms, manage R scripts and R objects, and integrate R results into applications or dashboards via SQL.

ORE is the top-level package for OML4R. The names of all packages in OML4R start with *ORE*. You can use the functionalities to drop a table (ore.drop) or create a table from a data frame (ore.create). You can connect to a database or disconnect from a database (ore.connect/ore.disconnect), check if you are connected (ore.is.connected), or list all the tables in the database (ore.ls). The transparency layer transparently transforms R expressions to SQL queries for a database to execute. It returns the results of this operation as an R object that OML4R understands.

The OREbase and OREstats packages are in the transparency layer. The transparency layer supports in-database data exploration, data preparation, and data analysis. It uses ore.frames instead of data.frames. The overloaded R functions do what the library is expected to do and takes full advantage of all Oracle Database functionalities, including database parallelism, query optimization, column indexing, and partitioning. OREbase includes conversion functionalities (as.ore*) to convert R types to OML4R types. It also includes ore.vector, ore.character, ore.factor, ore.frame, and ore.matrix functions for datatype conversion.

OREdplyr is a package for data manipulation. It uses ore.frames instead of data.frames for in-database execution to avoid costly data movements. It provides a subset of dplyr functionalities and is used for data manipulation, grouping, aggregation, sampling, and ranking.

Other packages include OREgraphics for graphics, OREeda for exploratory data analysis, OREdm for algorithms, OREmodels for models, OREpredict for scoring the models, and OREstats for statistical computations. To store R objects, use ore.save. To load them for an R session, use ore.load.

These are only examples of the packages for OML4R. There are plenty of other packages and functionalities.

With OML4R, you can write your own algorithms or install existing packages in database server-side R engines.

Using embedded R execution, you can execute R scripts in the database machine instead of your computer. You can invoke R from SQL or PL/SQL and return the results in Oracle tables, store and manage user-defined R functions in the database in the R script repository, and schedule user-defined R functions for automatic execution.

Python

Python can be used with an Oracle Database using cx_Oracle library, Oracle Cloud Infrastructure (OCI) Data Science, or OML4Py. Like OML4R, OML4Py is in the Oracle Database and executes the Python code on a database.

The rest of the solutions need to export the data from the database to use it. On the other hand, adding Python libraries is easier when using the database with cx_Oracle or OCI Data Science.

OML4Py can be used in Oracle Machine Learning Notebooks using the %python command. The main available libraries are cx_Oracle, warnings, Pandas, NumPy, Matplotlib, scikit-learn, sklearn, SciPy, and their dependencies. The OML4Py library is oml. To connect to a certain database, use oml.connect, and to specify the case table, use oml.sync (casetable=oml.sync(table = "EMPLOYEE")). You can create a table using oml.create.

```
oml.create(beer_df, table="BEERS")
```

You can access this table with SQL, Python, or PL/SQL. You can drop a table using oml.drop.

```
oml.drop(table="BEERS")
```

The OML library also includes the AutoML library. OML4Py AutoML automatically performs algorithm selection, feature selection, and hyperparameter tuning. To select the best algorithm, define the mining function and the scoring metric used, and the number of algorithms to compare. For instance, if you want to find the three best classification algorithms based on accuracy, you might write something like the following.

```
%python
as= automl.AlgorithmSelection(mining_function='classification',
score_metric='accuracy', parallel=2)
alg_ranking = as.select(Beer_data, Beer_rating, k=3)
print("Algorithms:\n", alg_ranking)
Algorithms:
 [('svm_gaussian', 0.9828757905208144), ('nn', 0.9710382092892445),
('svm_linear', 0.9449124812946643)]
```

This means the support-vector machines are the most accurate algorithm for the use case.

Similarly, automl.FeatureSelection can find the most meaningful features. automl. ModelTuning tunes a model and the hyperparameters. automl.ModelSelection skips these and automatically selects the best algorithm, and then builds, tunes, and returns the model.

You can use the Python script repository to store your own Python scripts in the database and grant privileges to other users to use your scripts with oml.grant. The following is an example.

```
%python
oml.grant(name="My_function", typ="pyqscript", user="OTHERUSER")
```

And you can revoke the privileges using oml.revoke.

```
%python
oml.revoke(name="My_function", typ="pyqscript", user="OTHERUSER")
```

There are plenty of machine learning process examples in Oracle Machine Learning Notebooks if you want to learn more.

Java

In September 2020, Oracle announced the availability of Tribuo (`https://tribuo.org`), as an open source Java Machine Learning library (`https://github.com/oracle/tribuo`).

This was an important announcement since many enterprise systems were built using Java. Adding machine learning functionalities that use other programming languages may not be the optimal choice. Tribuo makes it possible to build machine learning functionalities using Java. The algorithms offered are limited, but Tribuo also provides interfaces to ONNX Runtime, TensorFlow, and XGBoost. This means that if a model was stored in ONNX format or trained using TensorFlow or XGBoost, it can be deployed like any native Tribuo model.

OCI Data Science

Data Science is found in the OCI main menu. It is a payable option in Oracle Cloud and is not included in the free-tier offering.

The first step is to configure the Data Science tenancy. You can either configure it manually or using Oracle Resource Manager. To configure it manually, you need to do the following.

1. Create user groups and users for the data scientists.

2. Create compartments for network resources and Data Science resources.

3. Create the VCN and subnets.

4. Create policies to control access to network and Data Science-related resources.

5. Create dynamic groups and policies.

You can find detailed instructions in the OCI manual.

The other option is to use the Oracle Resource Manager and a predefined sample Data Science solution. First, create a compartment for Data Science work. Then open the OCI navigation menu. Go to Solutions and Platform ➤ Resource Manager and click Stacks, as shown in Figure 2-1.

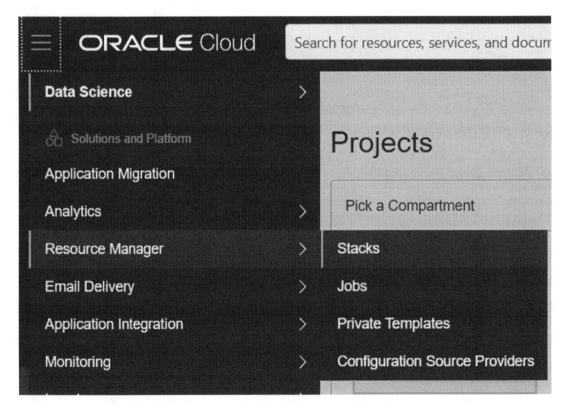

Figure 2-1. *Selecting stacks from the Resource Manager*

Next, select Create Stack ➤ Template ➤ Data Science (DS2). Click Select Template, as shown in Figure 2-2.

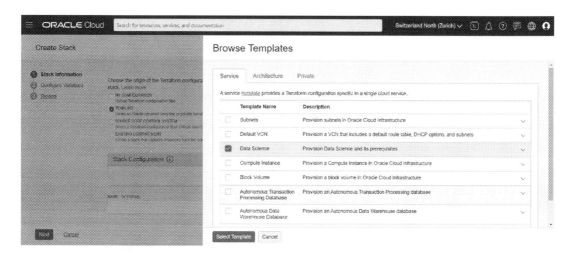

Figure 2-2. *Select the correct template from the list of templates.*

Check the information and make changes if needed. Click Next ➤ Next ➤ Create.
The stack has been created. To deploy the resources defined by this stack, go to
Terraform Actions and select Apply, as shown in Figure 2-3.

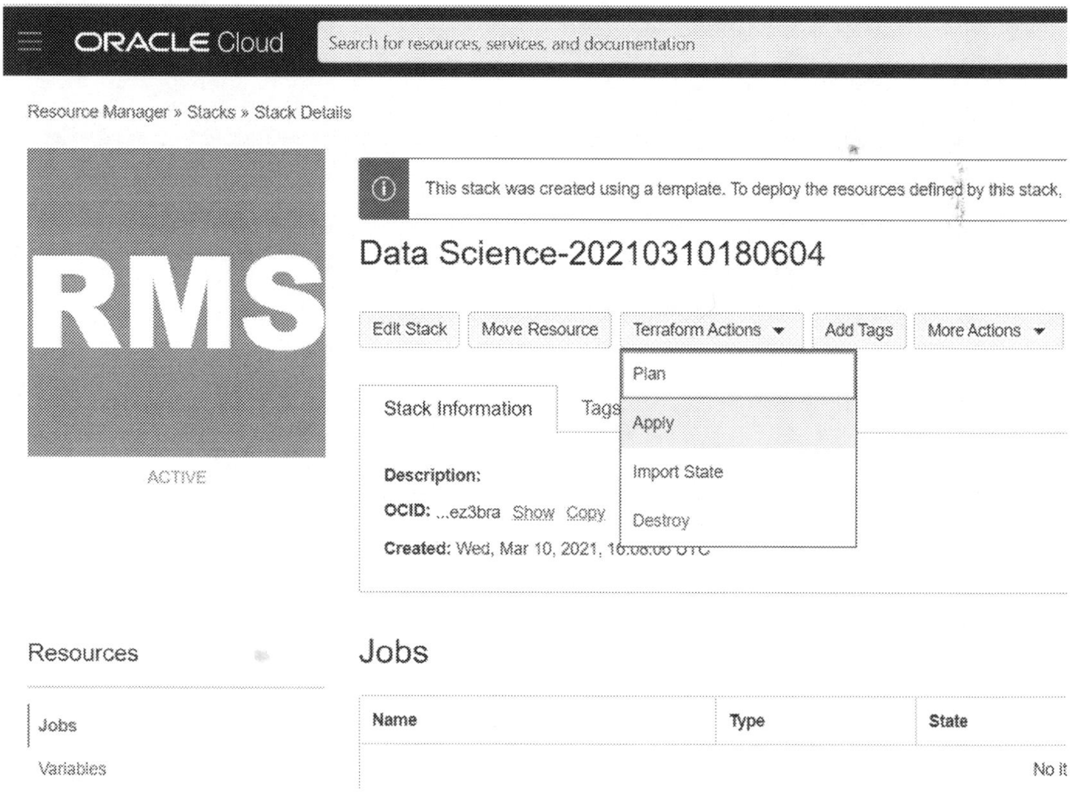

Figure 2-3. *Apply Terraform Actions*

Now the environment has been configured, and you can start working.

The work is done in Projects with Notebook Sessions. The next step is to create a project. From the OCI main menu, select Data Science ➤ Projects. Select the compartment you want to create the project in, and click Create Project. Enter the required information, and click Create. Next, you need to create a notebook session. Select the project, and click Create Notebook Session (see Figure 2-4).

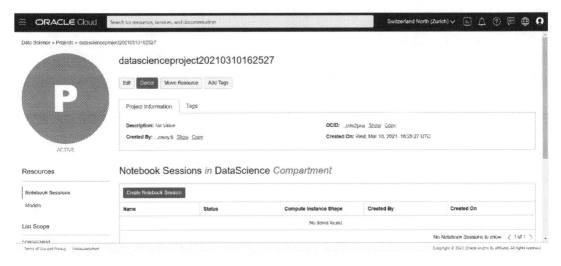

Figure 2-4. *Create Notebook Session*

Open a notebook session by clicking the Open button (see Figure 2-5). You can deactivate the notebook session with the Deactivate button and terminate it with the Terminate button. To terminate deletes everything, so you should first make a backup of the boot volume and attached block volume. Before deactivating a notebook session, make sure to save all your work to the attached block volume. Any data or files stored in the compute boot volume or held in memory are deleted when the notebook session is deactivated. The data and files saved on the block volume remain, and access is restored when the notebook session is activated.

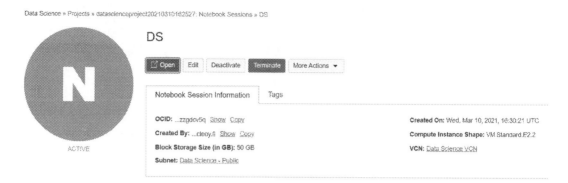

Figure 2-5. *Operating with a notebook session*

Figure 2-6 shows an example of a notebook session. The Oracle JupyterLab environment is pre-installed. It also includes default storage options for reading from and writing to OCI Object Storage. You can easily read from and write to several other data sources and data formats.

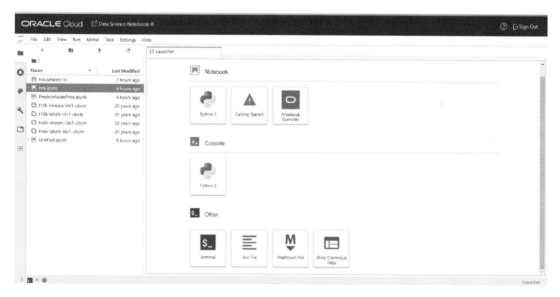

Figure 2-6. *A notebook session in OCI Data Science*

The Notebook works just like any notebook, and most of the Python libraries needed are already there. If you cannot find a library you want, you can add it yourself, just like in any Python environment, using the `pip install` command.

On top of a regular Python environment, you also have the Oracle Accelerated Data Science (ADS) SDK Python library included in the OCI Data Science service. It offers a friendly user interface, with objects and methods that cover all the steps involved in the life cycle of machine learning models, from data acquisition to model evaluation and interpretation. ADS SDK has automatic data visualization (`auto_transform`) for feature engineering and ADSEvaluator for model evaluations. AutoML automates feature selection, algorithm selection, feature encoding, and hyperparameter tuning.

OCI Data Science could be the best option for machine learning tasks on images and video, or if you already have a lot of code built using Python, or if you want to use ADS SDK and be able to install any Python libraries.

Oracle Analytics Cloud

Oracle Analytics Cloud (OAC) might be the best option for a business user or somebody who wants to analyze data without programming. It also has extremely powerful data preparation tools and the ability to use almost any data sources. You can use those data sources as single data, or you can use data flows in OAC to join data from different data sources as one and store it to the Oracle Database to be used with OML4SQL, OML4R, or OML4Py.

Oracle has three product lines for analytics: OAC, Oracle Analytics Server, and Oracle Analytics for Applications. Oracle Analytics makes preparing data much easier, offering augmented analytics and visualization. In OAC, you can use machine learning models created using OML4SQL. We talk more about OAC in Chapter 7.

AutoML

AutoML a very popular extension to many machine learning tools since it makes the model-building phase more efficient while making it easier for new users to start with machine learning.

The Oracle Autonomous database has a special AutoML functionality under Machine Learning. In Figure 2-7, you can see where to find it. This functionality is built using the same Python AutoML libraries mentioned earlier in this chapter.

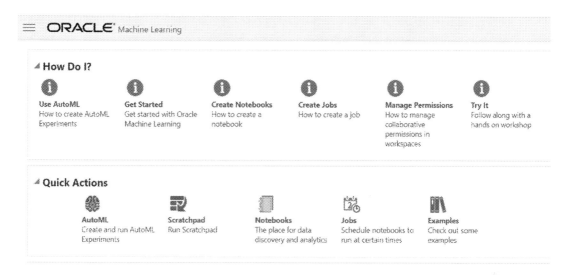

Figure 2-7. *AutoML in Oracle Machine Learning*

When you choose AutoML, the AutoML Experiments tool opens (see Figure 2-8). You can create new machine learning experiments or edit, delete, or duplicate existing ones. To start an experiment, click Start.

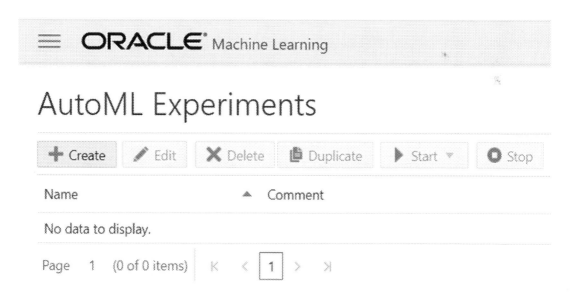

Figure 2-8. *AutoML experiments*

If you select Create, the Create Experiment window opens (see Figure 2-9). Here, you name the experiment and define the data source, target column, case ID, and prediction type. You can also select and deselect features as needed. When you are ready, select Start. There are two options: Faster Results and Better Accuracy. Choose the one you prefer and start the experiment process.

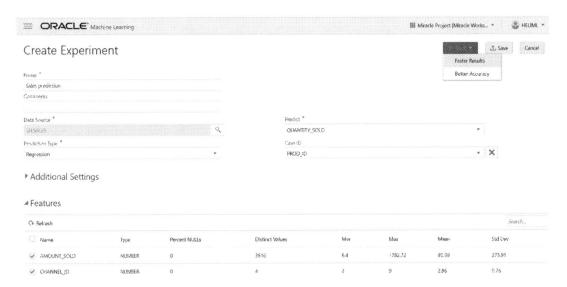

Figure 2-9. *Configuring the experiment process*

Running the experiment process takes a while. The results in Figure 2-10 show which algorithm performed the best (Leader Board) and how each of the features affected the result (Features).

Note   This is a dummy example. The idea is to show how AutoML can be used, not to show the machine learning process.

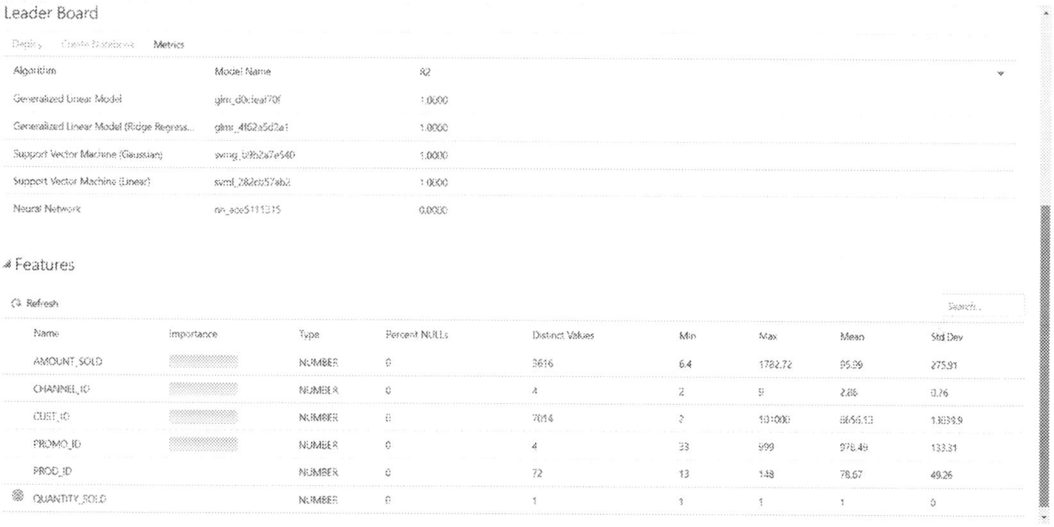

Leader Board

Algorithm	Model Name	R2						
Generalized Linear Model	glm_d0cleal70f	1.0000						
Generalized Linear Model (Ridge Regress...	glm_4f62a5d3a1	1.0000						
Support Vector Machine (Gaussian)	svmg_b9b2a7e54b	1.0000						
Support Vector Machine (Linear)	svml_282cb57ab2	1.0000						
Neural Network	nn_aoe5111315	0.0000						

Features

Name	Importance	Type	Percent NULLs	Distinct Values	Min	Max	Mean	Std Dev
AMOUNT_SOLD		NUMBER	0	3616	6.4	1782.72	95.39	275.91
CHANNEL_ID		NUMBER	0	4	2	9	2.86	0.76
CUST_ID		NUMBER	0	7014	2	101000	8656.53	13638.9
PROMO_ID		NUMBER	0	4	33	999	978.49	133.31
PROD_ID		NUMBER	0	72	13	148	78.67	49.26
QUANTITY_SOLD		NUMBER	0	1	1	1	1	0

Figure 2-10. *The leader board and feature analyses for the experiment*

R2, or accuracy, is used to compare the models, but you can use other metrics by selecting the metrics you want. Figure 2-11 shows other metrics that might be useful in that particular case.

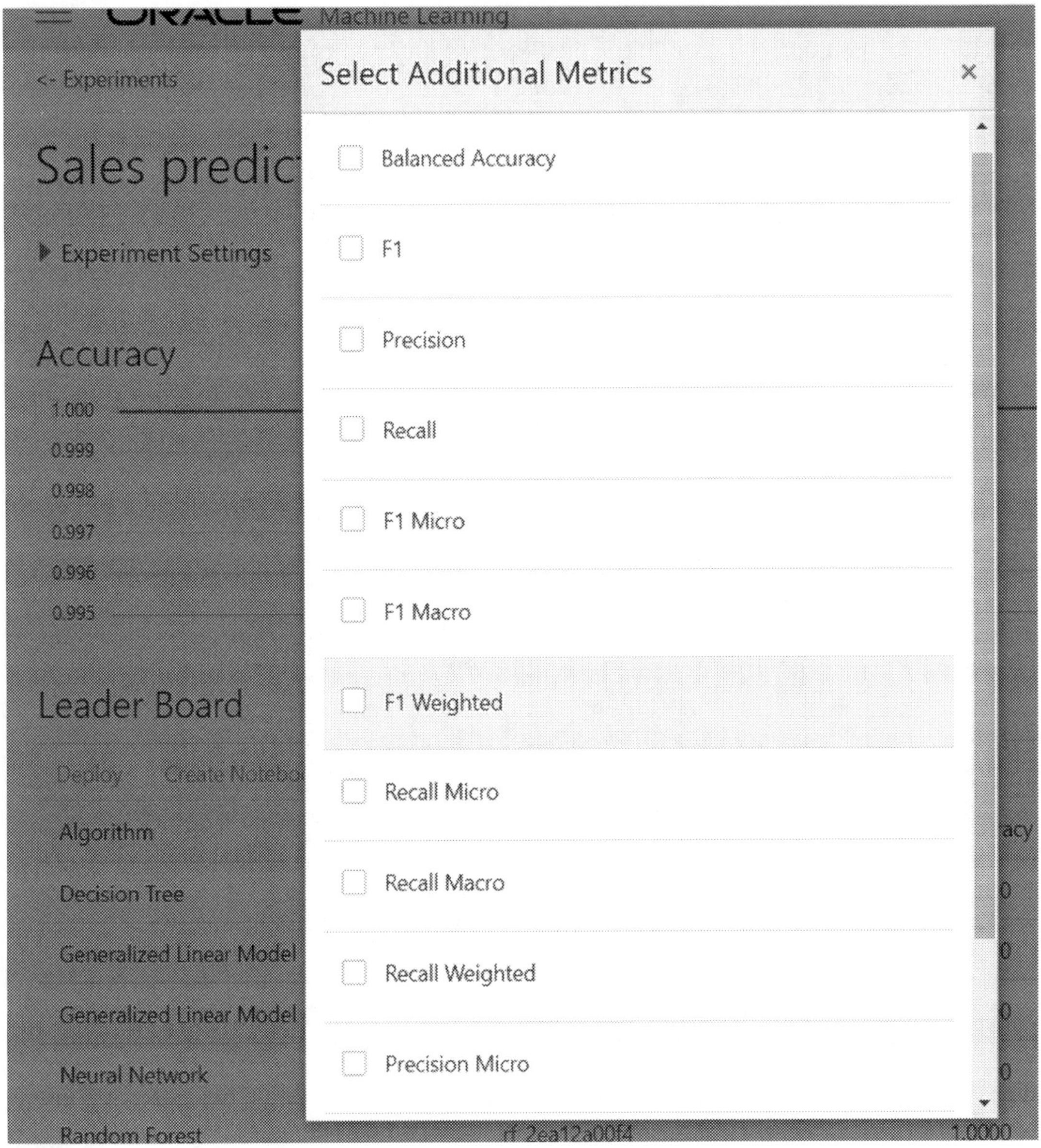

Figure 2-11. *Examples of other metrics to compare models*

You can select any of these models to see the prediction impact and confusion matrix. When you have decided on the best model, you can select it and create a notebook from it by selecting Create Notebook. Name the notebook and click OK.

Now this notebook can be seen and selected from the list of notebooks, just like any notebook. Figure 2-12 shows a part of the notebook created in the example.

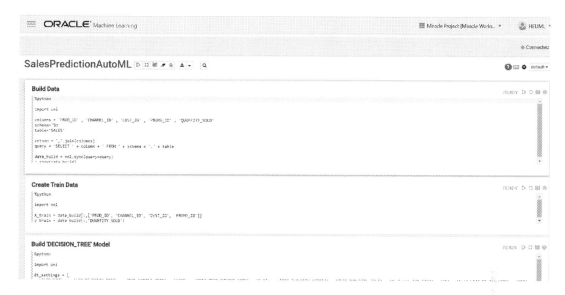

Figure 2-12. *A notebook created with AutoML*

Summary

Oracle has several machine learning offerings. You can use Oracle Analytics Cloud for preparing data and machine learning without the need to program. You can use OCI Data Science with Jupyter Notebooks and Python with the special Oracle ADS SDK machine learning library, including AutoML. Or you can use PL/SQL, R, or Python in the database without the need to move data outside the database. You can use several programming languages with interfaces to Oracle; for example, Python with cx_Oracle or Java with Tribuo. AutoML is popular because it makes building optimal machine learning models easier. Oracle offers AutoML in all its machine learning products.

CHAPTER 3

Oracle Machine Learning for SQL

Oracle Advanced Analytics consists of Oracle Data Mining and Oracle R Enterprise. On December 5, 2019, Oracle announced Advanced Analytics functionality as a cost-free option. Now, Oracle Data Mining has been rebranded to Oracle Machine Learning for SQL (OML4SQL), Oracle R Enterprise (ORE) to Oracle Machine Learning for R (OML4R) and Oracle R Advanced Analytics for Hadoop (OML4Spark).

New members of the OML family were also introduced. Oracle Machine Learning for Python (OML4Py), OML Microservices for Oracle applications, and Oracle Machine Learning Notebooks, Apache Zeppelin–inspired notebooks in Oracle Autonomous Database. This chapter discusses OML4SQL.

OML4SQL is a set of PL/SQL packages for machine learning implemented in Oracle Database. It is very convenient for Oracle professionals because all the knowledge they have gathered during the years working on Oracle technologies is still valid and very useful when learning new skills on machine learning. In OML4SQL, data is stored in a database or external tables, and the machine learning algorithms are inside the database. Instead of moving the data to be processed for machine learning, it stays where it is, and the processing is done there.

Moving data is often time-consuming and complicated. And what would be a better place to process data than a database. Since the data stays in the database, the security, backups, and so forth are managed by the database management system. The security risks related to moving data have been eliminated.

Having machine learning elements such as the model as a first-class citizen of the database makes handling them easy and similar to the handling process of any other database object. For instance, models can be trained in parallel. They can be exported from the test database and imported to the production database, and privileges can be defined easily and fine-grained.

© Heli Helskyaho, Jean Yu, Kai Yu 2021
H. Helskyaho et al., *Machine Learning for Oracle Database Professionals*,
https://doi.org/10.1007/978-1-4842-7032-5_3

Being able to do machine learning in a database brings a lot of advantages and fastens the process. However, the disadvantages are that the selection of algorithms depends on the database version, and adding new algorithms might be considered complicated. Though a simple solution usually yields the best result and existing algorithms are enough: good quality data, quick experiments, feature engineering, and model tuning rather than applying fundamentally different algorithms brings the largest improvements.

In Oracle Machine Learning, a feature (e.g., name or age) is called an *attribute*. A method (e.g., classification or clustering) is called a *machine learning function*. OML4SQL supports supervised and unsupervised machine learning. It supports neural networks and text processing, but it does not support convolutional neural networks and visual recognition problems.

PL/SQL Packages for OML4SQL

OML4SQL consists of three PL/SQL packages.

- DBMS_PREDICTIVE_ANALYTICS routines for performing semi-automized machine learning using predictive analytics

- DBMS_DATA_MINING_TRANSFORMING routines for transforming the data for OML4SQL algorithms

- DBMS_DATA_MINING routines for creating and managing the machine learning models and evaluating them

All of them have several subprograms, which are discussed later in this chapter.

Privileges

The three OML4SQL packages are owned by the SYS user. They are part of the database installation. The installation also includes granting the execution privilege on the packages to the public. When the routines in the package are run, it's done with those privileges the current user has (invoker's privileges). To use OML4SQL in an Oracle Autonomous Database, create the user using the Manage Oracle ML Users functionality in Administration functionality under Service Console. Oracle automatically adds the necessary privileges.

If you want to use OML4SQL in a non-autonomous database, you need to create the user and grant the required privileges manually. Users most likely need these system privileges because many steps in the machine learning process involve creating tables or views.

- CREATE SESSION

- CREATE TABLE

- CREATE VIEW

If the user needs the privilege to create machine learning models in his/her own schema, the CREATE MINING MODEL system privilege must be granted. That allows managing the models in his/her schema as well. If the models need to be created in other schemas, the CREATE ANY MINING MODEL system privilege is required. In that case, the following privileges should probably also be granted.

- ALTER ANY MINING MODEL

- DROP ANY MINING MODEL

- SELECT ANY MINING MODEL

- COMMENT ANY MINING MODEL

- AUDIT ANY MINING MODEL

If the user needs to do processing with text type data, EXECUTE ON ctxsys.ctx_ddl privilege is needed because OML4SQL uses Oracle Text functionalities when processing text.

It is also possible to grant individual object privileges on existing machine learning models. Designing the privileges is an important step of the process, and it is possible the privileges in the test environment will be different from those in the production environment.

Data Dictionary Views

Since all the OML4SQL objects are database objects, there are also USER_, DBA_, and ALL_ database views for their data as for any object in an Oracle database. USER_ views include objects owned by the user. DBA_ views are those the DBA-group members have access to. ALL_ views include all the views the user has privileges to. There are six of these views in OML4SQL.

- _MINING_MODELS

- _MINING_MODEL_ATTRIBUTES

- _MINING_MODEL_PARTITIONS

- _MINING_MODEL_SETTINGS

- _MINING_MODEL_VIEWS

- _MINING_MODEL_XFORMS

Let's use the USER_ view set as an example to investigate these views. USER_MINING_MODELS view has all the information about models created by the user who has logged in to the database. It includes the name of the model, what function was used, what algorithm was used, the algorithm type, creation date, duration of building the model, size of the model, is the model partitioned or not, and some optional comments about the model.

The view USER_MINING_MODEL_ATTRIBUTES describes all the attributes used when building the model. It includes information about the name of the model, name of the attribute, type of the attribute (categorical/numeric), datatype, length, precision, scale, usage type, information on whether this attribute is the target, and optional specifications for the attribute.

The view USER_MINING_MODEL_SETTINGS includes all the settings defined for models. The view includes the name of the model, the setting name and value, and a setting type indicating whether it is defined by the user with a settings table (INPUT) or is the default value (DEFAULT). The settings tables are discussed later in this chapter.

USER_MINING_MODEL_VIEWS includes information about all the views the machine learning process has created. These views are called *model detail views*. These views have been in a database since Oracle Database 12.2. They are for inspecting a model's details. USER_MINING_MODEL_VIEWS includes the name of the model, the name of the view, and the view type. The name of the view always starts with DM. The view types include Scoring Cost Matrix, Model Build Alerts, Classification Targets, Computed Settings, and Decision Tree Hierarchy. The views created depend on the algorithm.

USER_MINING_MODEL_XFORMS view describes the user-specified transformations embedded models. This is discussed later in this chapter.

USER_MINING_MODEL_PARTITIONS includes data about a partitioned model, which is covered later in this chapter.

Predictive Analytics

Predictive analytics is an easy option for an Oracle professional to use machine learning without creating models and understanding the process of machine learning in detail. Predictive analytics uses OML4SQL technology, but knowledge of machine learning processes is unnecessary since it automates parts of the process. The backbone of predictive analytics is a PL/SQL package called DBMS_PREDICTIVE_ANALYTICS that consists of three procedures.

- EXPLAIN

- PREDICT

- PROFILE

EXPLAIN ranks attributes in order of their influence in explaining the target column; in other words, it calculates the attribute importance. PREDICT predicts the value of a target column using the values in the input data. PROFILE generates rules that describe the cases from the input data. All mining activities are handled internally by these procedures, but at least some of the rows in the data set must have data on the target column.

The syntax of the EXPLAIN procedure is this.

```
DBMS_PREDICTIVE_ANALYTICS.EXPLAIN (
    data_table_name     IN VARCHAR2,
    explain_column_name IN VARCHAR2,
    result_table_name   IN VARCHAR2,
    data_schema_name    IN VARCHAR2 DEFAULT NULL);
```

The data_table_name refers to the table or view where the data is stored. Explain_column_name is the name of the target column. Result_table_name is the name of the created table and where all the data explaining the attributes is stored. If you want, you can also define the schema's name where the table/view is located, and the result table is created (data_schema_name). The default for the schema is the current schema. In the Beer_data table, there is data about beers. Let's look at how the columns in that table affect the beer's overall rating.

```
BEGIN
    DBMS_PREDICTIVE_ANALYTICS.EXPLAIN(
        data_table_name      => 'Beer_data',
        explain_column_name  => 'overall',
        result_table_name    => 'explain_overall');
END;
/
```

The result table created looks like this.

```
desc explain_overall
Name                 Null?  Type
----------------     -----  --------------
ATTRIBUTE_NAME              VARCHAR2(4000)
ATTRIBUTE_SUBNAME           VARCHAR2(4000)
EXPLANATORY_VALUE           NUMBER
RANK                        NUMBER
```

ATTRIBUTE_NAME refers to the data attribute name. ATTRIBUTE_SUBNAME refers to the model attribute name. If the data and the model attribute name are the same, the attribute subname is null. EXPLANATORY_VALUE describes how useful the attribute is in predicting the target value. The explanatory value is between 0 and 1. The bigger the value is, the more important the column is for prediction. The rank tells the order of the attributes in their importance, 1 being the most important. Let's see how important the attributes were in defining the overall rating.

```
select attribute_name, explanatory_value from explain_overall order by rank;
```

```
ATTRIBUTE_NAME       EXPLANATORY_VALUE
TASTE                0,26747389082648847849935989784339761753811
PALATE               0,19769621383921109816634703773225501265441
AROMA                0,15266215846201924233256539058672935183441
APPEARANCE           0,10013070459875254970399166918966898832011
IDINDEX              0,08426119156218495132520591358786053893861
BEERID               0,06587627900508279000106848865118805748991
STYLE                0,06351804213647736626012089128487555474221
BREWERID             0,05366510239972202444019009982181903827611
ABV                  0,03772335392602956900487292316950402451491
```

```
TIMEUNIX              0,0022278139762722427826044451470112412173
NAME                  0
AGEINSECONDS          0
GENDER                0
...
```

Ten of the columns have some importance in predicting the target attribute. Eight of them have no importance at all. If you carefully look at the results, you see that the most important columns (taste, palate, aroma, appearance) are null in a new data set (they are also evaluation results), so they cannot be directly used to predict the overall evaluation.

You could build a model and use a new data set, or use another technique, or accept that you cannot use them and remove the columns from the data set used for prediction. We discuss these steps later in this book.

One important step is to make sure there are not too many columns for building the prediction since each of them slows down the process and might end up in a less accurate prediction. In our example, we decided to remove the columns that we could not use by creating a view Beer_view.

```
CREATE VIEW beer_view AS
            SELECT IDINDEX, STYLE, BREWERID, ABV, OVERALL
            FROM Beer_data;
```

We kept columns taste, palate, aroma, and appearance since there is data in them. At this point, a beer expert would notice that the data is not the right kind of data for predicting the overall ratings because none of the attributes left would define the overall rating of a beer. But, since we have no business knowledge (big mistake!) we continued.

The next step is to use PREDICT to predict the target. The syntax to predict is as follows.

```
DBMS_PREDICTIVE_ANALYTICS.PREDICT (
    accuracy                OUT NUMBER,
    data_table_name         IN VARCHAR2,
    case_id_column_name     IN VARCHAR2,
    target_column_name      IN VARCHAR2,
    result_table_name       IN VARCHAR2,
    data_schema_name        IN VARCHAR2 DEFAULT NULL);
```

Accuracy returns the predictive confidence of the predicted value. data_table_name is the name of the table/view where the data is stored. case_id_column_name is the column name that identifies a case in the table/view. target_column_name is the name of the target column we are trying to predict. result_table_name is the name of the table that is created and where the result data stored. data_schema_name is the name of the schema where the source table/view is located and the result table is created.

```
DECLARE
  p_accuracy NUMBER(10,9);
BEGIN
  DBMS_PREDICTIVE_ANALYTICS.PREDICT(
        accuracy                => p_accuracy,
        data_table_name         =>'Beer_view',
        case_id_column_name     =>'idindex',
        target_column_name      =>'overall',
        result_table_name       => 'Predict_overall');
  DBMS_OUTPUT.PUT_LINE('Accuracy: ' || p_accuracy);
END;
/
```

The result table looks like this.

```
Name           Null? Type
-----------    ----- -------------
IDINDEX              NUMBER(7)
PREDICTION           NUMBER
PROBABILITY          BINARY_DOUBLE
```

For each IDINDEX (the caseID), it predicts the target column (overall) and the probability if it is a classification problem. For a regression problem, the probability is always null.

IDINDEX	PREDICTION	PROBABILITY
...		
6	4	0,6489473478441403
7	3	0,5399133550907291
9	4	0,4870673836213656
10	5	0,3541555965260737
11	4	0,623283467936882

12	4	0,36192909036295434
15	2	0,5106388569341178
16	4	0,5355598365975454
17	5	0,5772967022852917
19	4	0,44194795494972944

...

0.054057039

A prediction is returned for all cases, regardless of whether it had a value in the target column. If you want to compare the prediction to the actual value, you can compare those cases with an actual value to their predictions.

Note   It would be very smart to avoid column names like PREDICTION or PROBABILITY. Reserved words like these might cause problems when using functionalities.

The PROFILE procedure creates rules (expressed in XML as if-then-else statements) that describe the decisions that affected the prediction. PROFILE does not support nested types or dates. PROFILE generates a set of rules describing the expected outcomes, in our example values 0–5. Each profile includes a rule, record count, and score distribution. The syntax is as follows.

```
DBMS_PREDICTIVE_ANALYTICS.PROFILE (
       data_table_name          IN VARCHAR2,
       target_column_name       IN VARCHAR2,
       result_table_name        IN VARCHAR2,
       data_schema_name         IN VARCHAR2 DEFAULT NULL);
```

Let's try it to our data set.

```
BEGIN
    DBMS_PREDICTIVE_ANALYTICS.PROFILE(
          DATA_TABLE_NAME    => 'Beer_view',
          TARGET_COLUMN_NAME => 'overall',
          RESULT_TABLE_NAME  => 'Overall_profile');
END;
/
```

The result table it created (Overall_profile) looks like this.

```
Name            Null? Type
------------    ----- -----------
PROFILE_ID            NUMBER
RECORD_COUNT          NUMBER
DESCRIPTION           SYS.XMLTYPE

PROFILE_ID    RECORD_COUNT       DESCRIPTION
1             410                (XMLTYPE)
2             606                (XMLTYPE)
3             477                (XMLTYPE)
4             3337               (XMLTYPE)
...
22            63                 (XMLTYPE)
```

Data Preparation and Transformations

There are different interfaces for OML4SQL, but regardless of what you choose, it all starts with understanding the business needs and defining the task. Then you need to find the data that supports it. When the task has been defined, and the data is available, the next step is to prepare the data for the machine learning process. There are at least two reasons for that: you want to make sure the data is of good quality, and both OML4SQL and machine learning algorithms have some requirements for the data.

Understanding the Data

The process starts with describing, understanding, exploring, and analyzing the data. Since the data is in a database, there are several tools for that. You can use statistical functions, analytic functions, aggregation functions, or perhaps the procedures in the DBMS_STATS_FUNC package.

- EXPONENTIAL_DIST_FIT tests how well a sample of values fits an exponential distribution.

- NORMAL_DIST_FIT tests how well a sample of values fits a normal distribution.

- POISSON_DIST_FIT tests how well a sample of values fits a Poisson distribution.

- SUMMARY summarizes a numerical column of a table.

- UNIFORM_DIST_FIT tests how well a sample of values fits a uniform distribution.

- WEIBULL_DIST_FIT tests how well a sample of values fits a Weibull distribution.

This example studies the Student_enrollment table's AGE column with the SUMMARY procedure.

```
DECLARE
  summary_values DBMS_STAT_FUNCS.SummaryType;
  significance number;
BEGIN
  DBMS_STAT_FUNCS.SUMMARY(
    p_ownername => 'HELIML',
    p_tablename => 'STUDENT_ENROLLMENT',
    p_columnname => 'AGE',
    p_sigma_value => 3,
    s => summary_values);

  dbms_output.put_line('Summary statistics: ');
  dbms_output.put_line('Number of records: '
    ||summary_values.count);
  dbms_output.put_line('Min value: '||summary_values.min);
  dbms_output.put_line('Max value: '||summary_values.max);
  dbms_output.put_line('Variance: '
    ||round(summary_values.variance));
  dbms_output.put_line('Stddev: '||round(summary_values.stddev));
  dbms_output.put_line('Mean: '||summary_values.mean);
  dbms_output.put_line('Mode: '||summary_values.cmode(1));
  dbms_output.put_line('Median: '||summary_values.median);
```

```
  dbms_output.put_line('Quantiles');
  dbms_output.put_line('1st Quantile: '
    ||summary_values.quantile_5);
  dbms_output.put_line('2nd Quantile: '
    ||summary_values.quantile_25);
  dbms_output.put_line('3rd Quantile: '
    ||summary_values.quantile_75);
  dbms_output.put_line('4th Quantile: '
    ||summary_values.quantile_95);

  dbms_output.put_line('Extreme count: '
    ||summary_values.extreme_values.count);

  dbms_output.put_line('Top Five Values: '
    ||summary_values.top_5_values(1)||','
    ||summary_values.top_5_values(2)||','
    ||summary_values.top_5_values(3)||','
    ||summary_values.top_5_values(4)||','
    ||summary_values.top_5_values(5));

  dbms_output.put_line('Bottom Five Values: '
    ||summary_values.bottom_5_values(1)||','
    ||summary_values.bottom_5_values(2)||','
    ||summary_values.bottom_5_values(3)||','
    ||summary_values.bottom_5_values(4)||','
    ||summary_values.bottom_5_values(5));

  dbms_output.put_line('Normality test');
  DBMS_STAT_FUNCS.normal_dist_fit(
    ownername => 'HELIML',
    tablename => 'STUDENT_ENROLLMENT',
    columnname => 'AGE',
    test_type => 'SHAPIRO_WILKS',
    mean => summary_values.mean,
    stdev => summary_values.stddev,
    sig => significance);
END;
/
```

```
Summary statistics:
Number of records: 13848
Min value: 18
Max value: 55
Variance: 120
Stddev: 11
Mean: 36,4091565569035239745811669555170421721S
Mode: 18
Median: 36

Quantiles
1st Quantile: 19
2nd Quantile: 27
3rd Quantile: 46
4th Quantile: 53

Extreme count: 0
Top Five Values: 55,55,55,55,55
Bottom Five Values: 18,18,18,18,18

Normality test
W value : ,9538100006576519545941927251271296298S1
```

The best way for human eyes to understand the data is to visualize it. Several tools are available, including Oracle Machine Learning Notebooks, Oracle SQL Developer, Oracle Application Express (APEX), and Oracle Analytics Cloud.

Preparing the Data

The next step is preparing the data. Again, since the data is in a database, there are plenty of tools for that. The data preparation can be done using the enormous data processing functionalities and capabilities of an Oracle Database and/or using the functionalities in OML4SQL: Automatic Data Preparation (ADP) and a PL/SQL package called DBMS_DATA_MINING_TRANSFORM.

OML4SQL can only process data from one table or one view. A row in this table/view is called a *case*; therefore, the table or view is called a *case table*. Each record must be stored in a separate row and optionally identified by a unique case ID. Preparing the case table might include converting datatypes for some of the columns, creating nested columns, handling transactional data, or text transformations.

First of all, it is important to identify the columns to include in the case table. One of the steps is narrowing down the number of columns in the case table to those that are needed. For example, feature selection and feature extraction models can find the important attributes or replace attributes with better ones. For supervised learning, you need to define a specific attribute: the target. That is the attribute whose value we are trying to predict. Those attributes that are used for prediction are called *predictors*. A machine learning algorithm learns from the underlying data of these predictors and provides the prediction for a target.

The datatypes allowed for the target attribute are VARCHAR2, CHAR, NUMBER, FLOAT, BINARY_DOUBLE, BINARY_FLOAT, and ORA_MINING_VARCHAR2_NT for classification and NUMBER, FLOAT, BINARY_DOUBLE, and BINARY_FLOAT for regression machine learning problems. There is a limitation related to a target attribute: nested columns or columns of unstructured data (BFILE, CLOB, NCLOB, or BLOB) cannot be used as targets.

There are also two other kinds of attributes: data and model attributes. Data attributes are those columns in the data set that build, test, or score a model, whereas model attributes are the data representations used internally by the model. Data attributes and model attributes can be the same, but they can also be different. Data attributes can be any of the Oracle Database datatypes, but model attributes are either numerical, categorical, or unstructured (text).

Numerical attributes have an infinite number of values and an implicit order for both the values and the differences between the values. For OML4SQL datatypes NUMBER, FLOAT, BINARY_DOUBLE, BINARY_FLOAT, DM_NESTED_NUMERICALS, DM_NESTED_BINARY_DOUBLES, and DM_NESTED_BINARY_FLOATS are seen as numerical. For example, if an animal weighs 100 kg, it is heavier than the one that weighs 20 kg. If a column in a case table is defined as any of these datatypes, OML4SQL assumes the data is numeric and treats it that way. If the data is not numeric by its nature, then the datatype must be changed. For example, true/false expressed as 1/0 should not be defined as NUMBER since it can and should not be used for any comparisons (e.g., $0 < 1$) or calculus.

Categorical attributes have values that identify a finite number of discrete categories or classes. These categories have no order associated with the value. For example, if one person is 20 years old and another person is 50 years old, you can say the first person is younger, but you cannot say which person is "better." Instead, you can say that a person who is 51 years old is about the same age as the person who is 50 years old, so they belong in the same age group. In OML4SQL, categorical attributes are of type CHAR, VARCHAR2, or DM_NESTED_CATEGORICALS by default. These datatypes could also be of type unstructured text data. Datatypes CLOB, BLOB, and BFILE are always of type unstructured text data. Using several Oracle Text features, OML4SQL automatically transforms text columns so that the model can use the data. The text data must be in a table, not a view. The text in each row is treated as a separate document, and each document is transformed into a set of text tokens (terms), which have a numeric value and a text label. The text column is transformed to a nested column of DM_NESTED_ NUMERICALS.

If the data you want to use is not in one table, you need to combine the data from several tables to one case table (remember it can also be a view). If the relationship between two tables is one-to-many, such as one table for a customer and another table for the customer's different addresses, you need to use nested columns. To do that, you can create a view and cast the data to one of the nested object types. OML4SQL supports these nested object types: DM_NESTED_CATEGORICALS, DM_NESTED_ NUMERICALS, DM_NESTED_BINARY_DOUBLES, and DM_NESTED_BINARY_ FLOATS. Each row in the nested column consists of an attribute name/value pair, and OML4SQL internally processes each nested row as a separate attribute.

For example, there is data about the courses a student passed in the STUDENTS (studentid, firstname, lastname,...), COURSESPASSED (studentid, courseid, passeddate,...), and COURSES(courseid, coursename, credits,...) tables. If you created one table (a simplified table) based on them, it might look like the following.

```
select s.studentid, c.coursename, c.credits
from students s, courses c, coursespassed cp
where s.studentid=cp.studentid and
c.courseid=cp.courseid;
```

STUDENTID	COURSENAME	CREDITS
1	Database designing	5
1	Python programming	4
2	Database designing	5

```
2                Algorithms and logical thinking    6
3                Python programming                 4
...
```

Each student has several rows in the table, and the machine learning algorithm cannot use that kind of data.

Let's create a view (student_credits_view) based on that query.

```
create view student_credits_view as
(select s.studentid, c.coursename, c.credits
from students s, courses c, coursespassed cp
where s.studentid=cp.studentid and
c.courseid=cp.courseid);
```

Using that view, we transform the data to a column of type DM_NESTED_NUMERICALS in a students_credit_trans_nested view.

```
CREATE VIEW students_credit_trans_nested AS
SELECT studentid,
CAST(COLLECT(DM_NESTED_NUMERICAL
(coursename, credits)) AS DM_NESTED_NUMERICALS) studentcredits
FROM STUDENT_CREDITS_VIEW
      GROUP BY studentid;
```

This view can be used for OML4SQL because each student has only one row in the case table.

```
select * from students_credit_trans_nested;
STUDENTID    STUDENTCREDITS
1            SYS.DM_NESTED_NUMERICALS (SYS.DM_NESTED_NUMERICAL('Database
             designing', 5),
             SYS.DM_NESTED_NUMERICAL('Python programming', 4))
2            SYS.DM_NESTED_NUMERICALS (SYS.DM_NESTED_NUMERICAL('Database
             designing', 5),
             SYS.DM_NESTED_NUMERICAL('Algorithms and logical thinking', 6))
3            SYS.DM_NESTED_NUMERICALS
             (SYS.DM_NESTED_NUMERICAL('Python programming', 4))

...
```

OML4SQL supports Automatic Data Preparation (ADP), user-directed general data preparation, and user-specified embedded data preparation. Using the PREP_* settings in the Settings table described later in this chapter, you can control both automated or user-directed general data preparation. PREP_AUTO setting enables automated data preparation. The possible values are PREP_AUTO_ON and PREP_AUTO_OFF. PREP_AUTO_ON is the default. The other available PREP* settings require the PREP_AUTO being defined OFF to take effect. These settings are.

- PREP_SCALE_2DNUM, enables scaling data preparation for two-dimensional numeric columns. The following are possible values.

  - PREP_SCALE_STDDEV divides the column values by the standard deviation of the column. Combined with PREP_SHIFT_MEAN, the value yields to z-score normalization.

  - PREP_SCALE_RANGE divides the column values by the range of values. Combined with the PREP_SHIFT_MIN value, it yields a range of [0,1].

- PREP_SCALE_NNUM enables scaling data preparation for nested numeric columns. The following are possible values.

  - PREP_SCALE_MAXABS, yields data in the range of [–1,1].

- PREP_SHIFT_2DNUM enables centering data preparation for two-dimensional numeric columns. The following are possible values.

  - PREP_SHIFT_MEAN subtracts the average of the column from each value

  - PREP_SHIFT_MIN subtracts the column's minimum from each value

Some commonly needed transformations are feature scaling, binning, outlier treatment, and missing value treatment. These transformations are needed for algorithms to perform better. The need for a transformation depends on the algorithm used, data, and machine learning task. There is usually no need for feature scaling for tree-based algorithms, but it might make a big difference for gradient descent-based or distance-based algorithms.

In gradient descent-based algorithms (for example, linear regression, logistic regression, or neural network), the value of a feature affects the step size of the gradient descent. Since the difference in ranges of features causes different step sizes for each feature, the data is scaled before feeding it to the model to ensure that the gradient descent moves smoothly towards the minima. The distance-based algorithms (e.g., k-nearest neighbors, k-means, and support-vector machines) use distances between data points to determine their similarity. If different features have different scales, there is a chance that higher weightage is given to features with higher magnitude and cause a bias towards one feature. If you scale the data before employing a distance-based algorithm, the features contribute equally to the result, and most likely, the algorithm performs better.

Feature scaling can be divided into two categories: normalization and standardization. Normalization is usually a scaling technique where all the feature values end up ranging between 0 and 1: the minimum value in the column is set to 0, the maximum value to 1, and all the other values something between those.

Standardization centers the values around the mean with a unit standard deviation. Typically, it rescales data to have a mean of zero and a standard deviation of one (unit variance) using a z-score. Feature scaling eliminates the units of measurement for data, enabling you to more easily compare data from different domains. OML4SQL supports min-max normalization, z-score normalization, and scale normalization.

Feature scaling reduces the range of numerical data. Binning (discretization) reduces the cardinality of continuous and discrete data. Binning converts a continuous attribute into a discrete attribute by grouping values together in bins to reduce the number of distinct values. In other words, it transforms numerical variables into categorical counterparts. For instance, if the number of employees in the companies in our data set is 1–100000, you might want to bin them into groups for example 1–10, 11–50, 51–100, 101–1000, 1001–5000, 5001–8000, and more than 8000 employees. This is done because many machine learning algorithms only accept discrete values.

Binning can be unsupervised or supervised. Unsupervised binning might use quantile bins (25, 50, 75, 100), ranking, or it could be done with equal width or equal frequency algorithm. The equal width binning algorithm divides the data into k intervals of equal size with the width of w = (max-min)/k and interval boundaries: min+w, min+2w, ..., min+(k–1)w. Equal frequency/equal width binning algorithm divides the data into k groups, each containing approximately the same number of values. Supervised binning uses the data to determine the bin boundaries. Binning can improve resource utilization and model quality by strengthening the relationship between attributes. OML4SQL supports four different kinds of binning: supervised binning (categorical and numerical), top-n frequency categorical binning, equi-width numerical binning, and quantile numerical binning.

An *outlier* is a value that deviates significantly from most other values. Outliers can have a skewing effect on the data, and they can have a bad effect on normalization or binning. You need domain knowledge to determine outlier handling to recognize whether the outlier data is perfectly valid or problematic data caused by an error. Outlier treatment methods available in OML4SQL are trimming and winsorizing (clipping). Winsorizing sets the tail values of a particular attribute to some specified quantile of the data while trimming removes the tails by setting them to NULL.

Having NULLs in the data set can be problematic in many ways, but the data cannot be used for any machine learning activity since it does not exist. Rows with NULL values can be removed from the data set, or you can use missing value treatments. To make the decision, you must understand the data and the meaning of those NULLs for the business case. The automatic missing value treatment in OML4SQL has been divided into two categories: sparse data and data that contains random missing values.

In sparse data, missing values are assumed to be known, although they are not represented. Data missing in a random category means that some attribute values are unknown. For ADP, missing values in nested columns are interpreted as sparse data, and missing values in columns with a simple data type are assumed to be randomly missing. The treatment of a missing value depends on the algorithm and the data: categorical or numerical, sparse or missing at random. Oracle Machine Learning for SQL API Guide describes the missing value treatment for each case in detail. With some easy tricks, you can change the missing value treatment from the default. Replacing the nulls with a carefully chosen value (0, "NA"...) prevents ADP from interpreting it as missing at random, and therefore it does not perform missing value treatment. But be careful that it does not change the data too much. If you want missing nested attributes to be treated as missing at random, you can transform the nested rows into physical attributes. OML4SQL missing value treatment replaces NULL values for numerical datatypes with the mean and categorical datatypes with the mode.

Many machine learning algorithms have specific transformation requirements for the data, and ADP has predefined treatment rules for those algorithms supported in Oracle Database. When ADP is enabled, most algorithm-specific transformations are automatically embedded in the model during the model creation and automatically executed whenever the model is applied. It is also possible to change the rules, add user-specified transformations, or ignore ADP completely to perform all transformations manually. If the model has both ADP and user-specified transformations embedded, both sets of transformations are performed, but the user-specified transformations are performed first.

ADP does not affect the data set if the algorithm used is Apriori or decision tree. ADP normalizes numeric attributes if the algorithm is one of the following: generalized linear model (GLM), a k-means, non-negative matrix factorization (NMF), singular value decomposition (SVD), or support-vector machine (SVM). ADP bins all attributes with supervised binning if the algorithm is minimum description length (MDL) or naïve Bayes. If the algorithm is expectation maximization, ADP normalizes single-column numerical columns that are modeled with Gaussian distributions, but it does not affect the other types of columns. If the algorithm is an O-cluster, ADP bins numerical attributes with a specialized form of equal width binning and removes numerical columns with all nulls or a single value. If the algorithm is MSET-SPRT (multivariate state estimation technique–sequential probability ratio test), ADP uses z-score normalization. If you want to disable ADP for individual attributes, that can be done using a transformation list.

There are two ways of passing user-specified rules to CREATE_MODEL procedure when creating and training the model: passing a list of transformation expressions or passing the name of a view with transformations. When specified in a transformation list, the transformation expressions are executed by the model. When specified in a view, the transformation expressions are executed by the view. The main difference between these two options is that if you use a transformation list, the transformation expressions are embedded in the model and automatically implemented whenever the model is applied. But, if you use a view, the transformations must be re-created whenever applying the model. Therefore, the list approach is recommended over the view approach. The list can be build using CREATE*, INSERT*, and STACK* packages of DBMS_DATA_MINING_TRANSFORMATION, or they can be created manually. The views can be created using CREATE*, INSERT*, and XFORM* packages of DBMS_DATA_MINING_TRANSFORMATION, or they can be created manually.

If you want to use the DBMS_DATA_MINING_TRANSFORMATION package for the transformations, take the following steps.

1. Create the transform definition tables using appropriate CREATE* procedures. Each table has columns to hold the transformation definitions for a given type of transformation.

2. Define the transforms using the appropriate INSERT* procedures. It is also possible to populate them by yourself or modify the values in the transformation definition tables, but to do either of these, be sure you know what you are doing.

3. Generate transformation expressions. For that, there are two alternative ways.

    - STACK* procedures to add the transformation expressions to a transformation list to be passed to the CREATE_MODEL procedure and embedded into the model

    - XFORM* procedures to execute the transformation expressions within a view. These transformations are external to the model and need to be re-created whenever used with new data.

CREATE* procedures create the transformation definition tables for different kinds of transformations.

- CREATE_BIN_CAT for categorical binning

- CREATE_BIN_NUM for numerical binning

- CREATE_NORM_LIN for linear normalization

- CREATE_CLIP for clipping

- CREATE_COL_REM for column removal

- CREATE_MISS_CAT for categorical missing value treatment

- CREATE_MISS_NUM for numerical missing values treatment

The CREATE* procedures create transformation definition tables that include two columns, col and att. The column col holds the name of a data attribute and the name of the model attribute. If they are the same, the attribute is NULL, and the attribute name is simply col. If the data column is nested, then att holds the name of the nested attribute, and the name of the attribute is col.att. Neither the INSERT* nor the XFORM* procedures support nested data, and they ignore the att column in the definition table. The STACK* procedures and SET_TRANSFORM procedure support nested data. Nested data transformations are always embedded in the model.

INSERT* procedures insert definitions in a transformation definition table.

- INSERT_AUTOBIN_NUM_EQWIDTH: Numeric automatic equi-width binning definitions

- INSERT_BIN_CAT_FREQ: Categorical frequency-based binning definitions

- INSERT_BIN_NUM_EQWIDTH: Numeric equi-width binning definitions

- INSERT_BIN_NUM_QTILE: Numeric quantile binning definition

- INSERT_BIN_SUPER: Supervised binning definitions in numerical and categorical transformation definition tables

- INSERT_CLIP_TRIM_TAIL: Numerical trimming definitions

- INSERT_CLIP_WINSOR_TAIL: Numerical winsorizing definitions

- INSERT_MISS_CAT_MODE: categorical missing value treatment definitions

- INSERT_MISS_NUM_MEAN: numerical missing value treatment definitions

- INSERT_NORM_LIN_MINMAX: linear min-max normalization definitions

- INSERT_NORM_LIN_SCALE: linear scale normalization definitions

- INSERT_NORM_LIN_ZSCORE: linear z-score normalization definitions

STACK* procedures add expression to a transformation list.

- STACK_BIN_CAT: A categorical binning expression

- STACK_BIN_NUM: A numerical binning expression

- STACK_CLIP: A clipping expression

- STACK_COL_REM: A column removal expression

- STACK_MISS_CAT: A categorical missing value treatment expression

- STACK_MISS_NUM: A numerical missing value treatment expression

- STACK_NORM_LIN: A linear normalization expression

- DESCRIBE_STACK: Describes the transformation list

Each STACK* procedure call adds transformation records for all the attributes in a specified transformation definition table. The STACK* procedures automatically add the reverse transformation expression to the transformation list for normalization transformations. But they do not provide a mechanism for reversing binning, clipping, or missing value transformations.

If you want to use the views instead of transformation lists, the XFORM* procedures create the views of the data table.

- XFORM_BIN_CAT: Categorical binning transformations

- XFORM_BIN_NUM: Numerical binning transformations

- XFORM_CLIP: Clipping transformations

- XFORM_COL_REM: Column removal transformations

- XFORM_EXPR_NUM: Specified numeric transformations

- XFORM_EXPR_STR Specified categorical transformations

- XFORM_MISS_CAT: Categorical missing value treatment

- XFORM_MISS_NUM: Numerical missing value treatment

- XFORM_NORM_LIN: Linear normalization transformations

- XFORM_STACK: Creates a view of the transformation list

You can also create your own transformations and transformation lists without these procedures and definition tables. This approach gives more flexibility than using the procedures. Transformations for an attribute can be created using a record transformation (transform_rec). Each transform_rec specifies the transformation instructions for an attribute.

```
TYPE transform_rec IS RECORD (
    attribute_name       VARCHAR2(30),
    attribute_subname    VARCHAR2(4000),
    expression           EXPRESSION_REC,
    reverse_expression   EXPRESSION_REC,
    attribute_spec       VARCHAR2(4000));
```

The attribute whose value is transformed is identified by the attribute_name parameter or a combination of attribute_name.attribute_subname. The attribute_name refers to the data attribute, the column name. The attribute_subname refers to the name of a model attribute when the data attribute and the model attribute are not the same. Usually, that is in the case of nested data or text. If the attribute_subname is NULL, the attribute is identified by attribute_name. If it has a value, the attribute is identified by attribute_name.attribute_subname. You can specify a default nested transformation by setting NULL in the attribute_name field and the actual name of the nested column in the attribute_subname. You can also define different transformations for different attributes in a nested column. Using the keyword VALUE in the expression, define transformations based on the value of that nested column.

The expression defines the transformation for the attribute and the reverse_ expression, the reverse transformation for the transformation. Defining the reverse transformation is important for the model transparency and data usability since the transformed attributes might not be meaningful to the end users. They need to get the data transformed back to its original form. You can use the DBMS_DATA_MINING. ALTER_REVERSE_EXPRESSION procedure to specify or update reverse transformations expressions for an existing model.

The default value for attribute_spec is NULL. If ADP is not enabled in the model, the transformation record's attribute_spec field is ignored. If ADP is enabled, you can tell it not to touch the attribute value before your transformations are done (set the value to NOPREP), indicate the attribute is unstructured text (set the value to TEXT), or force a GML algorithm to include the attribute in the model build (set the value to FORCE_IN). If you use it to identify an attribute as unstructured text, it will not guide ADP behavior; but, it allows you to define other subsettings for the text column by using the following parameters.

- BIGRAM

- POLICY_NAME (Name of an Oracle Text policy object created with CTX_DDL.CREATE_POLICY)

- STEM_BIGRAM

- SYNONYM

- TOKEN_TYPE (NORMAL, STEM, or THEME)

- MAX_FEATURES

The following is an example.

```
"TEXT(POLICY_NAME:my_own_policy)(TOKEN_TYPE:THEME)(MAX_FEATURES:1500)"
```

You can read more about CTX_DDL and Oracle Text in Oracle manuals.

FORCE_IN forces the inclusion of the attribute in the model to build only if the ftr_selection_enable setting is enabled (the default is disabled), and it only works for the GLM algorithm. FORCE_IN cannot be specified for nested attributes or text. You can use multiple keywords by separating them with a comma (`"NOPREP, FORCE_IN "`).

With these transformation records, you can create a Transformation List as a collection of transformation records. This list can be created using the SET_TRANSFORM procedure in DBMS_DATA_MINING_TRANSFORM package and added to a model as a parameter when creating it. A transformation list is a stack of transformation records: when a new transformation record is added, it is appended to the top of the stack. Transformation lists are evaluated from bottom to top, and each transformation expression depends on the result of the transformation expression below it in the stack.

The process would look like this.

1. Write a SQL expression for transforming an attribute.

2. Write a SQL expression for reversing the transformation.

3. If you want to disable ADP for the attribute, define that.

4. Use the SET_TRANSFORM procedure to add these rules to a transformation list.

5. Repeat steps 1 through 4 for each attribute that you want to transform.

6. Pass the transformation list to the CREATE_MODEL procedure.

Each SET_TRANSFORM call adds a transformation record for a single attribute. SQL expressions specified with SET_TRANSFORM procedure must fit within a VARCHAR2 datatype. If the expression is longer than that, you should use the SET_EXPRESSION procedure that uses the VARCHAR2 array for storing the expressions. If you use SET_EXPRESSION to build a transformation expression, you must build a corresponding reverse transformation expression, create a transformation record, and add the transformation record to a transformation list. The GET_EXPRESSION function returns a row in the array.

It is worth noting that if an attribute is specified in the definition table and the transformation list, the STACK procedure stacks the transformation expression from the definition table on top of it in the transformation record and updates the reverse transformation.

PL/SQL API for OML4SQL

The DBMS_DATA_MINING PL/SQL package is the core of the OML4SQL machine learning. It contains routines for building, testing, and maintaining machine learning models, among others.

The Settings Table

When a model is created, all the parameters needed for training it are passed through a settings table. You should create a separate settings table for each model, insert the setting parameter values, and then assign the settings table to a model. If you do not attach a settings table to the model or define a setting, the default settings are used. There are global settings, machine learning function-specific settings, algorithm-specific settings, and settings for the data preparations.

All the setting tables used in model creations can be seen from USER_/ALL_/DBA_MINING_MODEL_SETTINGS database dictionary view. If a settings table has not been used, there are no rows in this view. It would be good practice to use naming conventions to identify which settings table belongs to which model. For the beer model using the decision tree algorithm, you might want to create a settings table named Beer_settings_DT or something similar that identifies it as a setting table for a beer data set for the decision tree algorithm.

Let's create three setting tables for different classification algorithms: decision tree (DT), naïve Bayes (NB), and support-vector machines (SVM).

```
CREATE TABLE Beer_settings_DT (
setting_name  VARCHAR2(30),
setting_value VARCHAR2(4000));

CREATE TABLE Beer_settings_NB (
  setting_name  VARCHAR2(30),
  setting_value VARCHAR2(4000));

CREATE TABLE Beer_settings_SVM (
  setting_name  VARCHAR2(30),
  setting_value VARCHAR2(4000));
```

The next step is to insert the setting wanted to that table. The insert clause always includes the name of the setting and the value. Since this is a decision tree model, you need to insert at least the algorithm information.

```
INSERT INTO Beer_settings_DT VALUES
  (dbms_data_mining.algo_name, dbms_data_mining.algo_decision_tree);
```

In this example, algo_name setting has an algo_decision_tree value. You add the data to the other two setting tables in the same way. The available algorithm names are listed in Table 3-1.

Table 3-1. *Algorithm Names for the Settings Table (Oracle Database 20c) (The default algorithm for each function is in italics)*

ML Function	ALGO_NAME Value (Name of the Algorithm)
Attribute importance, Feature selection	*ALGO_AI_MDL* (minimum description length) ALGO_CUR_DECOMPOSITION (CUR matrix decomposition) ALGO_GENERALIZED_LINEAR_MODEL (generalized linear model)
Association rules	ALGO_APRIORI_ASSOCIATION_RULES (Apriori)
Classification	ALGO_NAIVE_BAYES (naïve Bayes) ALGO_DECISION_TREE (decision tree) ALGO_GENERALIZED_LINEAR_MODEL (generalized linear model) ALGO_MSET_SPRT (multivariate state estimation technique–sequential) ALGO_SUPPORT_VECTOR_MACHINES (support-vector machine) ALGO_RANDOM_FOREST (random forest) ALGO_XGBOOST (XGBoost) ALGO_NEURAL_NETWORK (neural network)
Clustering	ALGO_KMEANS (enhanced k-means) ALGO_EXPECTATION_MAXIMIZATION (Expectation Maximization) ALGO_O_CLUSTER (O-Cluster)
Feature extraction	*ALGO_NONNEGATIVE_MATRIX_FACTOR* (non-negative matrix factorization) ALGO_EXPLICIT_SEMANTIC_ANALYS (explicit semantic analysis) ALGO_SINGULAR_VALUE_DECOMP (singular value decomposition)
Regression	ALGO_SUPPORT_VECTOR_MACHINES (support vector machine) ALGO_GENERALIZED_LINEAR_MODEL (generalized linear model) ALGO_XGBOOST (XGBoost)
Time series	ALGO_EXPONENTIAL_SMOOTHING (exponential smoothing)
All mining functions	ALGO_EXTENSIBLE_LANG (language used for extensible algorithms)

Each machine learning function also has a set of settings that can be defined. All the settings are described in the Oracle PL/SQL Packages and Types reference manual in the DBMS_DATA_MINING package documentation. Typically, these settings are *hyperparameters*, such as for clustering the maximum number of clusters (CLUS_NUM_CLUSTERS), and for association, the minimum confidence for association rules (ASSO_MIN_CONFIDENCE). Some of these settings are algorithm-specific. For example, for decision tree algorithm, there is a setting for the name of the cost table (CLAS_COST_TABLE_NAME) or for neural networks the Solver (optimizer) settings.

There are general settings that apply to any type of model but currently implemented only for specific algorithms. Those setting names start with ODMS_, and they include definitions for creating partitioned models, text processing, sampling, or random number seed. Using these general settings (ODMS_TABLESPACE_NAME), you can define the tablespace used for the model data. If this setting has not been provided, the default tablespace of the user is used.

There are also settings to configure the behavior of the machine learning model with an extensible algorithm (ALGO_EXTENSIBLE_LANG). When using an extensible algorithm, the model is built in R language. You can read more about it in the Oracle manual.

Model Management

Machine learning model management is easy: models can be created, renamed, and dropped. There are two procedures in DBMS_DATA_MINING package for creating a model: CREATE_MODEL and CREATE_MODEL2. CREATE_MODEL procedure creates a model based on data in a table or a view while as CREATE_MODEL2 creates it based on a query. A model can be renamed using RENAME_MODEL procedure, and dropped using DROP_MODEL procedure.

When creating a model using CREATE_MODEL you must define a name for the model, the mining function used, and the table/view name where the data is found for training the model. If you are building a prediction model, you also define the unique ID of the data and the target column name. The settings table and the data table schema name are not mandatory. The Xform_list parameter is for defining the data transformations discussed earlier in this chapter.

When the data and the attributes are ready for building the model, the next step is to choose the machine learning function used and see what algorithms exist for that function. Usually, the machine learning function is selected already when defining the task because that and the algorithms might define what kind of data transformations are needed. For instance, if the task is to predict whether a student will enroll for next semester, it is a classification problem. If the task is to predict the temperature in Helsinki next Monday, the problem is a type of regression.

OML4SQL offers several algorithms for different machine learning functions. Usually, every new version of Oracle Database introduces some new algorithms.

The CREATE_MODEL procedure in the DBMS_DATA_MINING package looks like this.

```
PROCEDURE CREATE_MODEL(
        model_name              IN VARCHAR2,
        mining_function         IN VARCHAR2,
        data_table_name         IN VARCHAR2,
        case_id_column_name     IN VARCHAR2,
        target_column_name      IN VARCHAR2 DEFAULT NULL,
        settings_table_name     IN VARCHAR2 DEFAULT NULL,
        data_schema_name        IN VARCHAR2 DEFAULT NULL,
        settings_schema_name    IN VARCHAR2 DEFAULT NULL,
        xform_list              IN TRANSFORM_LIST DEFAULT NULL);
```

Let's create a classification model with a decision tree algorithm (Beer_DT) for beer data set using the settings table created earlier (Beer_settings_DT). The original data set (Beer_data) is divided into two data sets: Beer_training_data and Beer_testing_data. And for building the model, training, the data set Beer_training_data is used. The other data set is left for testing the models.

Let's create a model using the training data and the decision tree setting table.

```
BEGIN
  DBMS_DATA_MINING.CREATE_MODEL(
    model_name              => 'Beer_DT',
    mining_function         => dbms_data_mining.classification,
    data_table_name         => 'Beer_training_data',
    case_id_column_name     => 'IDIndex',
    target_column_name      => 'Overall',
    settings_table_name     => 'Beer_settings_DT');
END;
/
```

The setting tables are created the same way as the models in the other two algorithms.

```
BEGIN
  DBMS_DATA_MINING.CREATE_MODEL(
    model_name          => 'Beer_NB',
    mining_function     => dbms_data_mining.classification,
    data_table_name     => 'Beer_training_data',
    case_id_column_name => 'IDIndex',
    target_column_name  => 'Overall',
    settings_table_name => 'Beer_settings_NB');
END;
/

BEGIN
  DBMS_DATA_MINING.CREATE_MODEL(
    model_name          => 'Beer_SVM',
    mining_function     => dbms_data_mining.classification,
    data_table_name     => 'Beer_training_data',
    case_id_column_name => 'IDIndex',
    target_column_name  => 'Overall',
    settings_table_name => 'Beer_settings_SVM');
END;
/
```

There are now three different models for predicting beer ratings: Beer_DT, Beer_NB, and Beer_SVM.

To rename a model, you simply call the RENAME_MODEL procedure with the current name of the model and the new name of the model.

```
EXEC dbms_data_mining.rename_model('Beer_DT_current', 'Beer_DT_new');
```

And to drop a model you call the DROP_MODEL procedure with the model name.

```
EXEC dbms_data_mining.drop_model('Beer_DT_new');
```

The models can be documented using a SQL Comment statement.

```
COMMENT ON MINING MODEL MLSchema.Beer_DT IS 'Decision Tree model predicts
beer overall rating';
```

If you want to remove the comment, you add an empty string (' ') as a comment. The data dictionary view USER_MINING_MODELS shows the comment as part of the model data in the Comments column. To comment on models created by other users, you need the COMMENT ANY MINING MODEL system privilege.

The Oracle auditing system can also track operations on machine learning models. To audit machine learning models, you must have the AUDIT_ADMIN role.

To know details about the model, you can query the data dictionary view USER_MINING_MODEL_VIEWS to see all the model detail views OML4SQL has created.

```
SELECT view_name, view_type FROM user_mining_model_views
WHERE model_name='BEER_DT'
ORDER BY view_name;
```

VIEW_NAME	VIEW_TYPE
DM$VCBEER_DT	Scoring Cost Matrix
DM$VGBEER_DT	Global Name-Value Pairs
DM$VIBEER_DT	Decision Tree Statistics
DM$VMBEER_DT	Decision Tree Build Cost Matrix
DM$VOBEER_DT	Decision Tree Nodes
DM$VPBEER_DT	Decision Tree Hierarchy
DM$VSBEER_DT	Computed Settings
DM$VTBEER_DT	Classification Targets
DM$VWBEER_DT	Model Build Alerts

Note   The GET_* procedures used for getting the information about model details in previous database versions are deprecated and replaced by model views.

Model Evaluation

When the model is built, it's time to evaluate how it works and which of the algorithms works the best for the task at hand. First, you use the APPLY procedure to apply the model to a new data set with known input and known output and create a result table.

```
BEGIN
    DBMS_DATA_MINING.APPLY(
        model_name          => 'Beer_DT',
        data_table_name     => 'Beer_testing_data',
```

```
        case_id_column_name => 'IDINDEX',
        result_table_name    => 'Beer_apply_results_DT',
        data_schema_name => 'ML');
 END;
/
```

The results table includes data for each case in the data set and the predictions and probabilities (including the possible cost).

After applying the new data set to the model, you can see the predictions and the probabilities in the results table. You can also use the RANK_APPLY procedure to see the top predictions for each case. The following is the syntax for that procedure.

```
DBMS_DATA_MINING.RANK_APPLY (
        apply_result_table_name          IN VARCHAR2,
        case_id_column_name              IN VARCHAR2,
        score_column_name                IN VARCHAR2,
        score_criterion_column_name      IN VARCHAR2,
        ranked_apply_table_name          IN VARCHAR2,
        top_N                            IN NUMBER (38) DEFAULT 1,
        cost_matrix_table_name           IN VARCHAR2    DEFAULT NULL,
        apply_result_schema_name         IN VARCHAR2    DEFAULT NULL,
        cost_matrix_schema_name          IN VARCHAR2    DEFAULT NULL);
```

By setting the value of the top_N parameter (1), you can get the most likely prediction or the top-n predictions for each case in the data set. The cost matrix information (cost_matrix_table_name, cost_matrix_schema_name) is only for classification. We discuss it later in this chapter.

The structure of apply_result_table_name depends on the machine learning function. The evaluation of the model created depends on the metrics and the machine learning function. For clustering, it looks like the following.

```
case_id        VARCHAR2/NUMBER,
cluster_id     NUMBER,
probability    NUMBER,
rank           INTEGER
```

And for classification, it looks like this.

```
case_id       VARCHAR2/NUMBER,
prediction    VARCHAR2/NUMBER,
probability   NUMBER,
cost          NUMBER,
rank          INTEGER
```

Often, the most important metric is accuracy. Accuracy means the portion of correct predictions of all the cases. The bigger the number, the more accurate the model is. But there are some limitations with using just accuracy for evaluating the models. What if both models are equally accurate, but one never predicts any overall beer rating to class 4? What if an overall rating of 1 instead of 5 is a bigger mistake than predicting an overall 4 instead of 5? This is why there are also other metrics for comparing the model performs better.

A confusion matrix compares predicted values to actual values building a matrix of four values: true positives (TP), false positives (FP), false negatives (FN), and true negatives (TN). The model's accuracy is defined as accurate predictions divided by the number of all cases (TP+TN)/(TP+TN+FP+FN).

DBMS_DATA_MINING package has a procedure to compute the confusion matrix automatically: COMPUTE_CONFUSION_MATRIX.

```
DBMS_DATA_MINING.COMPUTE_CONFUSION_MATRIX (
      accuracy                     OUT NUMBER,
      apply_result_table_name      IN  VARCHAR2,
      target_table_name            IN  VARCHAR2,
      case_id_column_name          IN  VARCHAR2,
      target_column_name           IN  VARCHAR2,
      confusion_matrix_table_name  IN  VARCHAR2,
      score_column_name            IN  VARCHAR2 DEFAULT 'PREDICTION',
      score_criterion_column_name  IN  VARCHAR2 DEFAULT 'PROBABILITY',
      cost_matrix_table_name       IN  VARCHAR2 DEFAULT NULL,
      apply_result_schema_name     IN  VARCHAR2 DEFAULT NULL,
      target_schema_name           IN  VARCHAR2 DEFAULT NULL,
      cost_matrix_schema_name      IN  VARCHAR2 DEFAULT NULL,
      score_criterion_type         IN  VARCHAR2 DEFAULT 'PROBABILITY');
```

In the syntax, you recognize case_id_column_name (the caseID), target_column_name, and the parameters for schemas for all the tables involved. target_table_name refers to a data set that has known input and output data, so the comparison between the prediction and the actual value can be made.

We used the training data set for building the model, and now it's time to use the testing data set to see how well the model performs. confusion_matrix_table_name is the new table this procedure creates. cost_matrix_table_name is optional and can be added if you have a cost table created with costs defined for misclassifications. We talk about the cost table a bit later.

score_criterion_type defines the scoring criteria used: probability or cost. The most interesting set of parameters is apply_result_table_name, score_column_name, and score_criterion_column_name. Before you can use the COMPUTE_CONFUSION_MATRIX procedure, you need to create the result table. You can create this table using the DBMS_DATA_MINING.APPLY procedure as we did earlier or SQL PREDICTION functions like this:

```
CREATE TABLE Beer_apply_results_DT AS
      SELECT IDINDEX,
      PREDICTION(Beer_DT USING *) prediction,
      PREDICTION_PROBABILITY(Beer_DT USING *)probability
      FROM Beer_testing_data;
```

Then you could use the result table like this to build the confusion matrix using probability as the score.

```
DECLARE
   v_accuracy      NUMBER;
BEGIN
   DBMS_DATA_MINING.COMPUTE_CONFUSION_MATRIX (
         accuracy                    => v_accuracy,
         apply_result_table_name     => 'Beer_apply_results_DT',
         target_table_name           => 'Beer_testing_data',
         case_id_column_name         => 'IDINDEX',
         target_column_name          => 'OVERALL',
         confusion_matrix_table_name => 'Beer_confusion_matrix_DT',
         score_column_name           => 'PREDICTION',
         score_criterion_column_name => 'PROBABILITY',
```

```
            cost_matrix_table_name       =>  null,
            apply_result_schema_name     =>  null,
            target_schema_name           =>  null,
            cost_matrix_schema_name      =>  null,
            score_criterion_type         =>  'PROBABILITY');
END;
/
```

A cost matrix can be used to affect the selection of a model by assigning costs or benefits to specific model outcomes. In OML4SQL the cost matrix is stored in a cost table. You create that table and then add the matrix data into that table.

Note   The actual and predicted target values datatypes must be the same as the model's target type.

Let's create a cost table and add data.

```
CREATE TABLE Beer_costs_DT (
  actual_target_value          NUMBER,
  predicted_target_value       NUMBER,
  cost                         NUMBER);

INSERT INTO Beer_costs_DT values (5, 5, 0);
INSERT INTO Beer_costs_DT values (5, 4, .25);
INSERT INTO Beer_costs_DT values (5, 3, .50);
INSERT INTO Beer_costs_DT values (5, 2, .75);
INSERT INTO Beer_costs_DT values (5, 1, 1);
INSERT INTO Beer_costs_DT values (4, 5, .25);
...
INSERT INTO Beer_costs_DT values (3, 1, .50);
INSERT INTO Beer_costs_DT values (2, 5, .75);
INSERT INTO Beer_costs_DT values (2, 4, .50);
INSERT INTO Beer_costs_DT values (2, 3, .25);
```

```
INSERT INTO Beer_costs_DT values (2, 2, 0);
INSERT INTO Beer_costs_DT values (2, 1, .25);
INSERT INTO Beer_costs_DT values (1, 1, 0);
...
COMMIT;
```

A cost matrix can be added to the confusion matrix using a parameter (cost_matrix_table_name) and changing the score_criterion_column_name and score_criterion_type to COST when creating the confusion matrix to use cost as the score.

```
DECLARE
    v_accuracy      NUMBER;
BEGIN
    DBMS_DATA_MINING.COMPUTE_CONFUSION_MATRIX (
            accuracy                     => v_accuracy,
            apply_result_table_name      => 'Beer_apply_results_DT',
            target_table_name            => 'Beer_testing_data',
            case_id_column_name          => 'IDINDEX',
            target_column_name           => 'OVERALL',
            confusion_matrix_table_name  => 'Beer_confusion_matrix_DT',
            score_column_name            => 'PREDICTION',
            score_criterion_column_name  => 'COST',
            cost_matrix_table_name       => null,
            apply_result_schema_name     => null,
            target_schema_name           => null,
            cost_matrix_schema_name      => null,
            score_criterion_type         => 'COST');
END;
/
```

Note   The data set used in building the confusion matrix or the cost matrix must be the same for building the result table.

You can embed the cost matrix to a classification model using the ADD_COST_MATRIX procedure, and you can remove a cost matrix using the procedure REMOVE_COST_MATRIX.

If the cost matrix has been assigned to a model, you can query the view
DM$VC*modelname*, (e.g., DM$VCBeer_DT) to see the cost matrix for the model.

For binary classification, there are two often used metrics: ROC and lift. The Receiver
Operating Characteristics (ROC) curve is a measure of how well a model can distinguish
between two classes. The ROC curve plots two quantities: recall (on the Y axis) and
specificity (on the X axis). A recall (also called the *true positive rate* or the *sensitive test*)
measures the portion of positives correctly identified: TP/(TP+FN). The specificity (also
called the *true negative rate*) defines how well the model defines the negative values:
TN/(TN+FT). The ROC curve measures the entire two-dimensional area, from (0,0) to (1,1),
underneath the entire ROC curve. That is called the Area Under the ROC Curve (AUC). To
compute the AUC of a ROC curve, use the COMPUTE_ROC procedure.

```
DBMS_DATA_MINING.COMPUTE_ROC (
        roc_area_under_curve            OUT NUMBER,
        apply_result_table_name         IN  VARCHAR2,
        target_table_name               IN  VARCHAR2,
        case_id_column_name             IN  VARCHAR2,
        target_column_name              IN  VARCHAR2,
        roc_table_name                  IN  VARCHAR2,
        positive_target_value           IN  VARCHAR2,
        score_column_name               IN  VARCHAR2 DEFAULT 'PREDICTION',
        score_criterion_column_name     IN  VARCHAR2 DEFAULT 'PROBABILITY',
        apply_result_schema_name        IN  VARCHAR2 DEFAULT NULL,
        target_schema_name              IN  VARCHAR2 DEFAULT NULL);
```

Let's test it with another data set. This data set has information about students and
the task is to predict whether the student will enroll for next semester or not.

```
DECLARE
    v_AUC       NUMBER;
BEGIN
    DBMS_DATA_MINING.COMPUTE_ROC (
        roc_area_under_curve     => v_AUC,
        apply_result_table_name  => 'Student_results_DT',
        target_table_name        => 'Student_testing_data',
        case_id_column_name      => 'Student_ID',
        target_column_name       => 'Enrolled',
```

```
        roc_table_name              =>  'Student_ROC_DT',
        positive_target_value       =>  '1',
        score_column_name           =>   'PREDICTION',
        score_criterion_column_name =>  'PROBABILITY');
END;
/
```

Note   If the target column is type NUMBER, instead of typing 1 into positive_target_value, type to_char(1).

The outcome is stored in a table the procedure creates (roc_table_name).

Name	Null?	Type
PROBABILITY		BINARY_DOUBLE
TRUE_POSITIVES		NUMBER
FALSE_NEGATIVES		NUMBER
FALSE_POSITIVES		NUMBER
TRUE_NEGATIVES		NUMBER
TRUE_POSITIVE_FRACTION		NUMBER
FALSE_POSITIVE_FRACTION		NUMBER

To measure how good a classification model is, you can use the gain and the lift charts. These charts measure how much better you could expect to do with the predictive model comparing without a model. The higher the lift, the better the model is.

The COMPUTE_LIFT procedure computes the lift for a classification model.

```
DBMS_DATA_MINING.COMPUTE_LIFT (
        apply_result_table_name      IN VARCHAR2,
        target_table_name            IN VARCHAR2,
        case_id_column_name          IN VARCHAR2,
        target_column_name           IN VARCHAR2,
        lift_table_name              IN VARCHAR2,
        positive_target_value        IN VARCHAR2,
        score_column_name            IN VARCHAR2 DEFAULT 'PREDICTION',
        score_criterion_column_name  IN VARCHAR2 DEFAULT 'PROBABILITY',
```

```
    num_quantiles                  IN NUMBER DEFAULT 10,
    cost_matrix_table_name         IN VARCHAR2 DEFAULT NULL,
    apply_result_schema_name       IN VARCHAR2 DEFAULT NULL,
    target_schema_name             IN VARCHAR2 DEFAULT NULL,
    cost_matrix_schema_name        IN VARCHAR2 DEFAULT NULL
    score_criterion_type           IN VARCHAR2 DEFAULT 'PROBABILITY');
```

The following is an example of how to use it with the student enrollment data.

```
BEGIN
    DBMS_DATA_MINING.COMPUTE_LIFT (
        apply_result_table_name       => 'Student_results_DT',
        target_table_name             => 'Student_testing_data',
        case_id_column_name           => 'Student_ID',
        target_column_name            => 'Enrolled',
        lift_table_name               => 'Student_Lift_DT',
        positive_target_value         => '1',
        score_column_name             => 'PREDICTION',
        score_criterion_column_name   => 'PROBABILITY',
        num_quantiles                 => 10,
        score_criterion_type          => 'PROBABILITY');
END;
/
```

The table created by the COMPUTE_LIFT looks like this.

```
Name                             Null? Type
------------------------------   ----- -------------
QUANTILE_NUMBER                        NUMBER
PROBABILITY_THRESHOLD                  BINARY_DOUBLE
GAIN_CUMULATIVE                        NUMBER
QUANTILE_TOTAL_COUNT                   NUMBER
QUANTILE_TARGET_COUNT                  NUMBER
PERCENTAGE_RECORDS_CUMULATIVE          NUMBER
LIFT_CUMULATIVE                        NUMBER
TARGET_DENSITY_CUMULATIVE              NUMBER
TARGETS_CUMULATIVE                     NUMBER
```

```
NON_TARGETS_CUMULATIVE              NUMBER
LIFT_QUANTILE                       NUMBER
TARGET_DENSITY                      NUMBER
```

Model Scoring and Deployment

Most machine learning models can be applied to new data in a process known as *scoring*. OML4SQL supports the scoring operation for classification, regression, anomaly detection, clustering, and feature extraction. Deployment refers to the use of models in a new environment. For instance, if the model was built in the test database, deployment copies the model to the production database to score production data.

Note   Data that is scored must be transformed in the same way as the data for training the model.

Scoring can be done to data in real time or in batches. It can include predictions, probabilities, rules, or some other statistics. Just like any SQL query in Oracle Database, OML4SQL scoring operations support parallel execution. If parallel execution is enabled, scoring operations can use multiple CPUs and I/O resources to get significant performance improvements. The scoring process matches column names in the scoring data with the names of the columns used when building the model. Not all the columns in model building need to be present in the scoring data. If the data types do not match, OML4SQL tries to convert the datatypes.

Earlier in this chapter, we said that the scoring data must undergo the same transformations as the corresponding column in the build data so that the model can evaluate it. If the transformation instructions are embedded in the model, transformations are done automatically. And if ADP is enabled, the transformations required by the algorithm are also performed automatically.

Scoring can be done using the APPLY or RANK_APPLY procedure, which creates results tables, or by using OML4SQL's special scoring functions that return the predictions as a query result set. The following creates predictions for a new data set (Beer_bryggeri) using the model Beer_DT and the APPLY procedure.

```
EXEC dbms_data_mining.apply('Beer_DT','Beer_bryggeri', 'IDINDEX',
'Bryggeri_Result_Table');

select * from Bryggeri_Result_Table;
```

IDINDEX	PREDICTION	PROBABILITY	COST
1	4	0,5437677813054055	0,4562322186945945
1	5	0,30132354760235847	0,6986764523976415
1	3	0,12155197295179977	0,8784480270482002
1	2	0,02824392858615429	0,9717560714138457
1	1	0,005071537541747413	0,9949284624582526
1	0	0,00004123201253453181	0,9999587679874654
2	4	0,5437677813054055	0,4562322186945945
2	5	0,30132354760235847	0,6986764523976415
2	3	0,12155197295179977	0,8784480270482002
2	2	0,02824392858615429	0,9717560714138457
2	1	0,005071537541747413	0,9949284624582526
2	0	0,00004123201253453181	0,9999587679874654
3	4	0,5437677813054055	0,4562322186945945
3	5	0,30132354760235847	0,6986764523976415
3	3	0,12155197295179977	0,8784480270482002
3	2	0,02824392858615429	0,9717560714138457

...

There are scoring functions for different kinds of machine learning functions, and the selection of a scoring function depends on the machine learning function. The prediction details, how the prediction is done, is stored in XML strings. There are functions to show these prediction details.

- CLUSTER_DETAILS returns the details of a clustering algorithm.

- FEATURE_DETAILS returns the details of the feature engineering algorithm.

- PREDICTION_DETAILS returns the details of a classification, regression, or anomaly detection algorithm.

These functions return the actual value of attributes used for scoring and the relative importance of the attributes in determining the score. By default, they return the five most important attributes for scoring.

The following functions can be used with a classification, regression, or anomaly detection algorithm for predictions.

- PREDICTION returns the best prediction for the target.

- PREDICTION_PROBABILITY returns the probability of the prediction.

- PREDICTION_COST returns the cost of incorrect predictions. PREDICTION_SET returns the results of a classification model, with the predictions and probabilities for each case.

- PREDICTION_BOUNDS (GLM only) returns the upper and lower bounds of the interval where the predicted values (linear regression) or probabilities (logistic regression) lie.

This example returns the prediction and the prediction details for a beer with IDINDEX=6 using model Beer_DT.

```
SELECT idindex, PREDICTION(Beer_DT USING *) pred,
PREDICTION_DETAILS(Beer_DT USING *) preddet
FROM Beer_bryggeri
WHERE idindex = 6;
```

Scoring functions use the USING clause to define which attributes are for scoring. If you use * instead of a list of attributes, all the attributes to build the model are used; but you can define only some of the attributes or define an expression.

This example uses all the attributes in the model to predict the overall rating of a beer with IDINDEX=6.

```
SELECT PREDICTION (Beer_DT USING *)
FROM beer_bryggeri where IDINDEX = 6;
```

This example uses only the STYLE attribute to predict the overall rating of a beer with IDINDEX=6.

```
SELECT PREDICTION (Beer_DT USING STYLE)
FROM beer_bryggeri where IDINDEX = 6;
```

This example would predict the overall rating for a beer having the word "Bryggeri" in the text column.

```
SELECT PREDICTION(Beer_DT USING 'Bryggeri' AS text)
FROM DUAL;
```

This example gives the probability of beer with IDINDEX=6 having an overall rating of 5, using all the attributes used on the model building.

```
SELECT PREDICTION_PROBABILITY(Beer_DT, 5 USING *) as beer_overall_prob
FROM beer_bryggeri
WHERE idindex = 6;
```

```
0,30132354760235847
```

If the model is partitioned (discussed later in this chapter), the ORA_DM_PARTITION_NAME procedure returns the name of the partition where the data locates. This example predicts student enrollment and in which partition (female or male) the result comes from.

```
SELECT prediction(STUDENT_PART_CLAS_SVM using *) pred, ora_dm_partition_
name(STUDENT_PART_CLAS_SVM USING *) partname FROM studentenrolment;
```

```
PRED    PARTNAME
1       Female
0       Male
0       Male
0       Female
1       Female
0       Male
0       Male
0       Male
...
```

Another specialty for a partitioned model is a GROUPING hint. It partitions the input data set to score each partition in its entirety before the next partition. Using this hint when scoring large data sets with several partitions might lead to better performance, but using it with small data or large data with just a few partitions might lead to worse performance. Here's an example of how to use the GROUPING hint.

```
SELECT PREDICTION(/*+ GROUPING */STUDENT_PART_CLAS_SVM USING *) pred
FROM studentenrolment;
```

There are also scoring functions for feature engineering.

- FEATURE_ID returns the ID of the feature and its highest coefficient value.

- FEATURE_COMPARE compares two similar and dissimilar sets of texts.

- FEATURE_SET returns a list of objects containing all possible features along with the associated coefficients.

- FEATURE_VALUE returns the value of the predicted feature.

The following describes the scoring functions for clustering algorithms.

- CLUSTER_ID returns the ID of the predicted cluster.

- CLUSTER_DISTANCE returns the distance from the centroid of the predicted cluster.

- CLUSTER_PROBABILITY returns the probability of a case belonging to a given cluster.

- CLUSTER_SET returns a list of all possible clusters to which a given case belongs, along with the associated probability of inclusion.

```
select idindex,
CLUSTER_ID(Beer_KMEANS USING *) as clus,
CLUSTER_PROBABILITY(Beer_KMEANS USING *) as prob,
CLUSTER_DISTANCE (Beer_KMEANS USING *) as dist
from Beer_bryggeri;
```

IDINDEX	CLUS	PROB	DIST
19281	8	0,3446464115268269	0,017476842283835947
5981	9	0,2813443932152596	0,2996019036949169
19177	8	0,36398701665384614	0,015375735728568318
9783	7	0,3920713116202475	0,00734828191959036
...			

To deploy the model to a new Oracle Database instance, you can use EXPORT_MODEL and IMPORT_MODEL procedures. The EXPORT_MODEL creates a dump file that the IMPORT_MODEL can read. To identify the location of the file, they use a directory object. A directory object is a logical name in the database for a physical directory on the host computer. First, you need to create the directory. For that, you need the CREATE ANY DIRECTORY privilege. To export machine learning models to that directory, you must have write access to the directory object and file system directory. You must have read access to the directory object and file system directory to import machine learning models. The database must also have access to the file system. Create the directory and grant privileges needed to the test database machine learning user and the production database machine learning user.

```
CREATE OR REPLACE DIRECTORY oml_model_dir AS '/dm_path/oml_models';

GRANT READ, WRITE ON DIRECTORY oml_model_dir TO mlusertest;
GRANT READ ON DIRECTORY oml_model_dir TO mlusetprod;
```

Connect to the test database as the test database machine learning user and export the model.

```
BEGIN
   dbms_data_mining.export_model (
      filename =>   'BeerDT.dmp',
      directory => 'oml_model_dir');
      model_filter => 'name in (''Beer_DT'')');
END;
/
```

Of course, you can also export all the models by omitting the model_filter parameter, or export all the models using decision tree algorithm by setting the parameter to 'ALGORITHM_NAME IN ("DECISION_TREE")', or all models using clustering by setting the parameter to 'FUNCTION_NAME = "CLUSTERING"'.

Here's an example of a IMPORT_MODEL procedure call.

```
BEGIN
   dbms_data_mining.import_model (
                   filename => 'BeerDT.dmp',
                   directory => 'oml_model_dir',
                   schema_remap => 'TestML:ProdML',
                   tablespace_remap => 'EXAMPLE:SYSAUX');
END;
/
```

From Oracle Database Release 18c onward, EXPORT_SERMODEL and IMPORT_SERMODEL procedures are available to export/import serialized models. The serialized format allows the models to be moved outside the database for scoring. The model is exported in a BLOB datatype and can be saved in a BFILE. The import procedure takes the serialized content in the BLOB and creates the model.

This example uses EXPORT_SERMODEL procedure to export a model (Beer_DT) in a serialized format and store it in a table (modelexpblob). You can save it to a BFILE to be able to share it outside the database.

```
DECLARE
   v_blob blob;
BEGIN
   dbms_lob.createtemporary(v_blob, FALSE);
   dbms_data_mining.export_sermodel(v_blob, 'Beer_DT');
   INSERT INTO modelexpblob (modeldesc) values (v_blob);
   dbms_lob.freetemporary(v_blob);
END;
/
```

Then import the model from serialized format to the database. In our example, the model is stored in a column of a table.

```
DECLARE
   v_blob blob;
BEGIN
   -- dbms_lob.createtemporary(v_blob, FALSE);
   SELECT modeldesc into v_blob from modelexpblob;
```

```
    -- you can also fill in v_blob from a file and then you need
    -- the other lines marked as comments
    dbms_data_mining.import_sermodel(v_blob, 'IMP_MODEL');
    -- dbms_lob.freetemporary(v_blob);
END;
/
```

Let's see what the model looks like from the USER_MINING_MODEL view.

```
SELECT model_name, mining_function, algorithm, algorithm_type
FROM user_mining_models WHERE model_name = 'IMP_MODEL';
```

```
MODEL_NAME    MINING_FUNCTION    ALGORITHM       ALGORITHM_TYPE
IMP_MODEL     CLASSIFICATION     DECISION_TREE   NATIVE
```

Predictive Model Markup Language (PMML) is an XML-based predictive model interchange format. If regression models were created in another environment, but you can export them into a PMML format, you can import them to Oracle Database using the IMPORT_MODEL procedure. The IMPORT_MODEL procedure is overloaded: you can call it to import mining models from a dump file set, or you can call it to import a single mining model from a PMML document.

The limitation is that the models must be of type RegressionModel, either linear regression or binary logistic regression.

This is the syntax for importing a mining model from a PMML document.

```
DBMS_DATA_MINING.IMPORT_MODEL (
    model_name      IN  VARCHAR2,
    pmmldoc         IN  XMLTYPE
    strict_check    IN  BOOLEAN DEFAULT FALSE);
```

For example, importing the PMML_regression model from the PMMDIR directory from the HousePrice_regression.xml file.

```
BEGIN
    dbms_data_mining.import_model ('PMML_regression',
    XMLType (bfilename ('PMMLDIR', 'HousePrice_regression.xml'),
            nls_charset_id ('AL32UTF8')));
END;
/
```

If you do not want to build a model, you can use dynamic scoring. Instead of supplying the predefined model to the OML4SQL function, you supply an analytic clause. The function builds one or more transient models and uses them to score the data. There are some limitations for dynamic scoring: the models created during dynamic scoring are not available for inspection or fine-tuning, such as selecting the algorithm or cost matrix. This example predicts the age of a student. It returns the student ID, real age, predicted age, the difference between age and predicted age, and the model's description.

```
SELECT student_id, age, pred_age, age-pred_age age_diff, pred_det
FROM (SELECT student_id, age, pred_age, pred_det,
    RANK() OVER (ORDER BY ABS(age-pred_age) DESC) rnk FROM
    (SELECT student_id, age,
        PREDICTION(FOR age USING *) OVER () pred_age,
        PREDICTION_DETAILS(FOR age ABS USING *) OVER () pred_det
  FROM studentenrolment))
WHERE rnk <= 5;
```

Partitioned Model

One of the benefits of using machine learning in a database is the possibility to partition the model. A partitioned model organizes and represents multiple models as partitions in a single model entity, enabling you to easily build and manage models tailored to independent slices of data. You must include the partition columns as part of the USING clause when scoring.

The partitioning is done using a partitioning key(s). If there are several columns in the key, it is represented as a comma-separated list of up to 16 columns. The partitioning key horizontally slices the input data based on discrete values of the partitioning key. The partitioning key must be either of type NUMBER or VARCHAR2.

A partitioned model is build using the CREATE_MODEL procedure with a settings table including the information about partitioning. During the model build process, the data is partitioned based on the distinct values of the partitioning key. The model partitions compose a model, but they cannot be used as standalone models. The maximum number of partitions is 1000. If you want, you can define a different maximum when creating the model using the ODMS_MAX_PARTITIONS setting in the settings table.

Let's create an SVM model for the student data and partition it by gender. First, you need to create a settings table and insert the data needed. Because this is a partitioned model, you need to add a row that defines the partitioning key (gender).

```
BEGIN
  INSERT INTO Student_part_settings VALUES
    (dbms_data_mining.algo_name,
     dbms_data_mining.algo_support_vector_machines);
  INSERT INTO Student_part_settings VALUES
    (dbms_data_mining.prep_auto, dbms_data_mining.prep_auto_on);
  INSERT INTO Student_part_settings VALUES
(dbms_data_mining.svms_kernel_function,dbms_data_mining.svms_linear);
  -- define that it will be partitioned by GENDER
  INSERT INTO Student_part_settings VALUES
    (dbms_data_mining.odms_partition_columns, 'GENDER');
  COMMIT;
END;
/
```

Then create the model using the settings table.

```
BEGIN
  DBMS_DATA_MINING.CREATE_MODEL(
    model_name          => 'Student_part_SVM',
    mining_function     => dbms_data_mining.classification,
    data_table_name     => 'Studentenrolment',
    case_id_column_name => 'student_id',
    target_column_name  => 'enrolled',
    settings_table_name => 'Student_part_settings');
END;
/
```

You can see the details of the model from the data dictionary view USER_MINING_MODEL_PARTITIONS.

```
select MODEL_NAME as model, PARTITION_NAME as partition, POSITION, COLUMN_
NAME as cname, COLUMN_VALUE as cvalue from ALL_MINING_MODEL_PARTITIONS;
```

MODEL	PARTITION	POSITION	CNAME	CVALUE
STUDENT_PART_SVM	Male	1	GENDER	Male
STUDENT_PART_SVM	Female	1	GENDER	Female

If you query the USER_MINING_MODEL_ATTRIBUTES view (remember to write the model's name with capital letters!), you can see that the GENDER attribute is of type PARTITION.

```
SELECT attribute_name, attribute_type
  FROM user_mining_model_attributes
 WHERE model_name = 'STUDENT_PART_SVM'
ORDER BY attribute_name;
```

ATTRIBUTE_NAME	ATTRIBUTE_TYPE
AGE	NUMERICAL
CREDITS	NUMERICAL
ENROLLED	CATEGORICAL
FIRSTNAME	CATEGORICAL
GENDER	PARTITION
LASTNAME	CATEGORICAL

The data dictionary view USER_MINING_MODEL_VIEWS shows the views created by OML4SQL for this model.

```
SELECT view_name, view_type FROM user_mining_model_views
WHERE model_name='STUDENT_PART_SVM'
ORDER BY view_name;
```

VIEW_NAME	VIEW_TYPE
DM$VCSTUDENT_PART_CLAS_SVM	Scoring Cost Matrix
DM$VGSTUDENT_PART_CLAS_SVM	Global Name-Value Pairs
DM$VLSTUDENT_PART_CLAS_SVM	SVM Linear Coefficients
DM$VNSTUDENT_PART_CLAS_SVM	Normalization and Missing Value Handling
DM$VSSTUDENT_PART_CLAS_SVM	Computed Settings
DM$VTSTUDENT_PART_CLAS_SVM	Classification Targets
DM$VWSTUDENT_PART_CLAS_SVM	Model Build Alerts

There are several procedures in the DBMS_DATA_MINING package for managing partitioned models.

- ADD_PARTITION adds partition/partitions in an existing partition model.

- DROP_PARTITION drops a partition.

- COMPUTE_CONFUSION_MATRIX_PART computes the confusion matrix for the evaluation of partitioned models.

- COMPUTE_ROC_PART computes ROC for evaluation of a partitioned model.

- COMPUTE_LIFT_PART computers lift for evaluation of partitioned models.

When adding a partition, you might end up in a situation where partition keys of the new partition conflict with those of the existing partitions. You have three options from which to choose how the conflict is solved.

- ERROR terminates the procedure without adding any partitions.

- REPLACE replaces the existing partition for which the conflicting keys are found.

- IGNORE eliminates the rows having conflicting keys.

If you want to drop the model, simply use DROP_MODEL procedure.

Extensions to OML4SQL

Because OML4SQL is in the database as PL/SQL packages, you can call those packages from any SQL tool, such as Oracle SQL Developer, SQLcl, or SQL*Plus. But there are also some special interfaces: Oracle Data Miner and OML Notebooks.

Oracle Data Miner and Oracle SQL Developer

Oracle Data Miner is an extension to Oracle SQL Developer. They are both available to download for free from Oracle Technology Network. Oracle Data Miner is a graphical interface for OML4SQL. To use it, you need to have Oracle SQL Developer installed.

You need to create a user for Oracle Data Miner with data mining privileges (something like listed here) and privileges to the tables/views where the data exists.

```
CREATE USER dmuser IDENTIFIED BY password
       DEFAULT TABLESPACE default_tablespace
       TEMPORARY TABLESPACE temp_tablespace
       QUOTA UNLIMITED on default_tablespace;
GRANT create mining model TO dmuser;
GRANT create procedure TO dmuser;
GRANT create session TO dmuser;
GRANT create table TO dmuser;
GRANT create sequence TO dmuser;
GRANT create view TO dmuser;
GRANT create job TO dmuser;
GRANT create type TO dmuser;
GRANT create synonym TO dmuser;
```

If you plan to use text data, you should also grant privileges to Oracle Text package (ctxsys.ctx_ddl).

```
GRANT EXECUTE ON ctxsys.ctx_ddl TO dmuser;
```

Then, you need to create a connection to the database. Select Tools, Data Miner, Make Visible from the Oracle SQL Developer menu to get the Data Miner connection pane visible in Oracle SQL Developer. Create a Data Miner connection to the database. Then double-click the connection in the Data Miner connections pane. If there is no Data Miner repository installed on the database, it suggests installing it. If you select Yes, the installation starts by asking for a password for the SYS user. This is a requirement. You cannot install the Data Miner repository without the SYS password. Explicitly, you cannot install the repository to the autonomous database because you do not have the password. In the autonomous database, you can use the OML Notebooks and their visualization capabilities. The repository creation process automatically creates the repository, and it is immediately ready to be used. If you need to remove the repository, you simply go to Tools ➤ Data Miner ➤ Drop Repository.

Oracle Data Miner is a drop-and-drag tool for Oracle Machine Learning. You start a new machine learning project by right-clicking the connection and selecting New Project (see Figure 3-1).

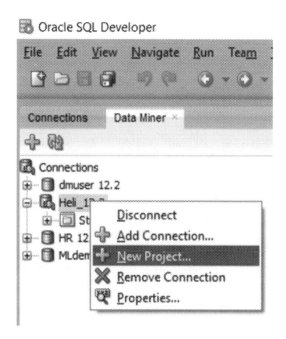

Figure 3-1. *Creating a project in Data Miner*

Then you give the project a name. Then right-click on the Project name and select New Workflow. Give a name to the Workflow. Then just drag-and-drop components from the Workflow Editor. You use Link from Linking Nodes to link different elements. Figure 3-2 shows Student_training_data as a data source. It is connected to Explore Data using a link defined in Explore Data. The data is grouped by the enrolled target attribute. Run Explore Data by right-clicking it and selecting Run. Right-click and select View Data to see the data. To see how the attributes correlate to the target attribute, select View Statistics.

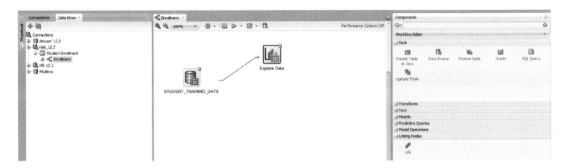

Figure 3-2. *Exploring data using Oracle SQL Developer and Data Miner*

Figure 3-3 shows whether age has any correlation to enrollment. Red indicates those who have not enrolled, and yellow indicates those who have enrolled. Based on this, there is no strong correlation between age and enrollment.

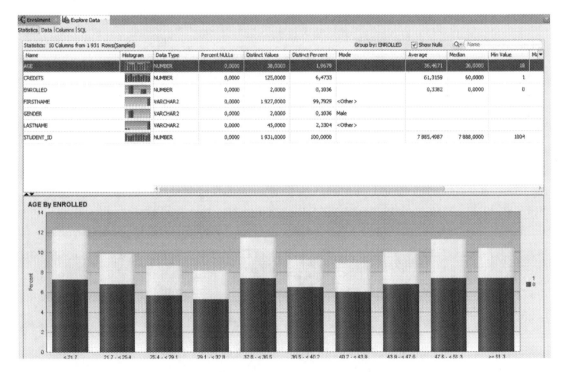

Figure 3-3. *Age by enrolled visualization*

When you are ready with the data, you can start building a model. Our target is to predict whether the student will enroll or not. Therefore, it is a classification problem and select classification from the Workflow Editor. Then define the target and the case ID, as shown in Figure 3-4.

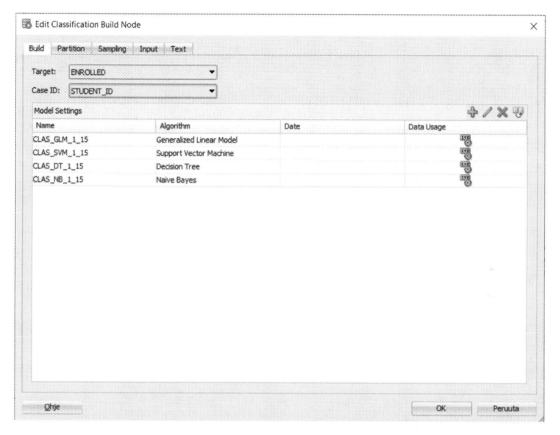

Figure 3-4. *Defining the model setting for classification model*

Next, run the model. It uses all the algorithms available in the database for classification problems. You can right-click the Class Builder icon and select View Models, View Test Results, or Compare Test Results.

Figure 3-5 shows the Compare Test Results output, in which Data Miner automatically compares the different algorithms. You can choose the one that is best for your purposes.

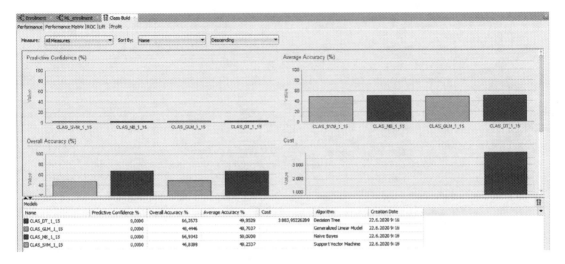

Figure 3-5. *Comparing the different algorithms using Compare Test Results*

After choosing the best algorithms, you can deselect the other algorithms in the Models pane. Add a new data set, and select Apply. Then, define the case ID. Link the new data set, Apply, and the Class Build. Then run Apply and get the predictions, as shown in Figure 3-6.

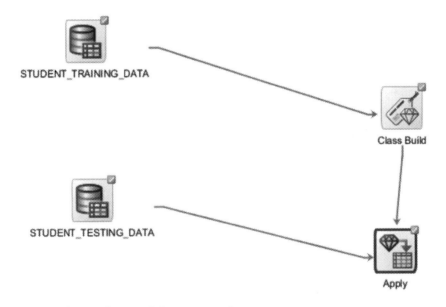

Figure 3-6. *Applying the model to a new data set*

Now right-click on the Apply and select View Data to see the predictions. Select Deploy to deploy the model.

Data Miner is very easy to use if you understand the machine learning process. There is no need for any coding. All is done by clicking and drag-and-dropping.

OML Notebooks

In Oracle Cloud, there is a possibility to use an autonomous database. There are two different workloads for an autonomous database: transaction processing (ATP) and data warehousing (ADW). Both of these include one more option to Oracle Machine Learning: OML Notebooks. OML Notebooks are Apache Zeppelin notebooks. Using these notebooks machine learning process is easy to document, and it includes many data visualization functionalities. The OML Notebooks are using the same PL/SQL packages discussed earlier in this chapter. You can read more about OML Notebooks and the autonomous database in Chapters 4, 5, and 6.

Summary

OML4SQL operates in Oracle Database. That means that there is no need to move the data for the machine learning process. It can be processed in the database. Also, the machine learning models are database objects and can be used as any database object. There are PL/SQL packages in the database to prepare and transform the data for machine learning and build and evaluate the machine learning models. Those packages can be used from any interface that allows calling PL/SQL packages. They can also be used with the GUI of Oracle Data Miner plug-in in Oracle SQL Developer or OML Notebooks in OCI.

CHAPTER 4

Oracle Autonomous Database for Machine Learning

The next three chapters explore machine learning with Oracle Cloud, specifically in the Oracle Autonomous Database environment, which offers many great benefits for machine learning projects. First, you can take advantage of various Oracle Cloud Infrastructure services such as compute, storage, and memory, as well as the Autonomous Database platform. These allow you to easily provide self-management and highly scalable machine learning development and runtime platform in the Cloud without spending time and resources on environment creation and management. As a result, you can focus more on your machine learning development. Second, you can leverage a comprehensive set of machine learning algorithms in Autonomous Database to meet various machine learning project needs. Third, you can use Oracle Machine Learning, a built-in tool that offers a collaborative web-based interactive data analysis environment for data scientists. This tool includes a SQL interface and data mining notebook designed for data analytics, data discovery, and data visualizations, as well as model building and evaluation for your machine learning project.

While the next two chapters focus on Oracle Machine Learning functionalities and how to use them for machine learning projects, this chapter describes the Oracle Autonomous Database environment, including provisioning and access, loading data, and importing a database schema. These are essential tasks and practices for developing machine learning projects. At the end of this chapter, we discuss how to access Oracle Machine Learning with Autonomous Database. This chapter covers the following topics.

© Heli Helskyaho, Jean Yu, Kai Yu 2021
H. Helskyaho et al., *Machine Learning for Oracle Database Professionals*,
https://doi.org/10.1007/978-1-4842-7032-5_4

- Oracle Cloud Infrastructure and Autonomous Database

- Oracle Autonomous Database architecture and components

- Working with Oracle Autonomous Database

- Oracle Machine Learning with Oracle Autonomous Database

Oracle Cloud Infrastructure and Autonomous Database

A quick way to start an Oracle Machine Learning project is to leverage the Oracle Autonomous Database. Oracle Cloud Infrastructure (OCI) offers three types of autonomous databases that are optimized for their corresponding workloads.

- **Autonomous Data Warehouse (ADW)**: Optimized for workloads like data warehouses, business analytics, and machine learning.

- **Autonomous Transaction Processing (ATP)**: Optimized for transaction processing and real-time analytical applications and mixed workloads such as transactions, batch, reporting, Internet of Things (IoT), application development, and so on.

- **Oracle Autonomous JSON Database**: A cloud document database optimized for developing JSON-centric applications.

While these autonomous databases share the common autonomous attributes and common cloud infrastructure, we still can highlight some of the major differences between them. For example, ADW uses the columnar data format, whereas ATP uses the row data format. ADW uses memory mainly for parallel joins and data aggregations, whereas ATP uses memory for data caching to reduce IOs for ATP. ADW aims to accelerate the data access for creating data summaries, whereas ATP focuses on remote direct memory access (RDMA) for messaging and IO. Since both ADW and ATP support the same Oracle Machine Learning environment, let's focus on ADW as the example in this chapter.

OCI offers all three types of Oracle autonomous databases. Let's review OCI first.

Oracle Cloud Infrastructure Services

Oracle Cloud Infrastructure provides a set of cloud services that you can leverage to build and run applications in a highly available environment. These services include a core cloud infrastructure and various database services and services for big data and artificial intelligence (AI). The core cloud infrastructure provides compute, storage, and networking services. One of the important database services that OCI provides is Oracle Autonomous Database, which consists of Autonomous Data Warehouse for business analytics and machine learning workloads, Autonomous Transaction Process for transactional type workloads, and the Autonomous JSON Database for JSON applications.

Through OCI, you can easily provision these infrastructure resources for developing and running a wide range of applications and services such as databases and business analytics applications without spending much time and effort on hardware and infrastructure installation and configuration. For example, when you request to provision an Autonomous Database instance, the provisioning process creates the following services in OCI.

- Compute resources such as virtual machine(s) with the required CPUs and memory for database servers to run database instances

- Network resource that creates and manages the network components for the database service (e.g., configures a virtual cloud network (VCN) to support the public and private network traffic to the database service)

- A storage service that stores all the database files of the database

In addition to these essential infrastructure services that run an Autonomous Database instance, you may use other OCI services, such as the Object Storage service, which provides high throughput storage for various types of data. For example, if you need to load data from an external source to Autonomous Database in Oracle Cloud, you must first upload the data from this external source to the Object Storage service in OCI. Then you can copy the data to your ADW instance.

Sign-up and Access Oracle Cloud Infrastructure

To start using Autonomous Database service or other services in OCI, you need to register an Oracle Cloud Free Tier account or a paid account. An Oracle Cloud Free Tier account provides a free trial with US$300 of free credit for up to 30 days plus Always Free services. Within the 30-day free trial period, you can use all the services in OCI until the free credits run out. To continue the account after the free trial ends, you can either upgrade to a paid account or remain with the Always Free services. The following Always Free services are available for an unlimited time with resource capacity restrictions on each service.

- Autonomous Data Warehouse

- Autonomous Transaction Processing

- Compute

- Block Volume

- Object Storage

- Archive Storage

- Load Balancing

- Monitoring

- Notifications

- Outbound Data Transfer

The following explains how to register for an Oracle Cloud Free Tier account in OCI.

1. Go to the Oracle Cloud Free Tier page at `www.oracle.com/cloud/free/`. Click the Start for Free button.

2. On the Oracle Cloud Free Tier signup screen, select your country/territory, enter your name and email address, and accept the Terms of Use.

3. This leads to the next screen (see Figure 4-1) which allows you to enter your cloud account information, including the cloud account name. Select the account type and the appropriate home region that is Always Free Eligible.

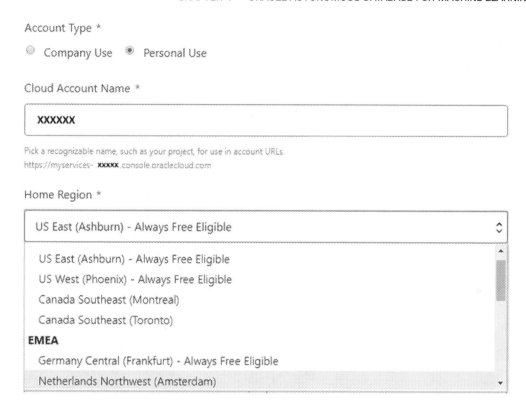

Figure 4-1. *Sign up for an Oracle Cloud Infrastructure account*

4. Enter your account password and mobile number to receive the verification code.

5. Provide your credit card payment information. You aren't charged unless you select to upgrade to a paid account.

Shortly after you sign up for an account, you receive a confirmation email from Oracle with the cloud account name, the username (your email address), and a link to your cloud account login page.

Figure 4-2 shows the main page after logging in to the Oracle Cloud. The Quick Action section shows shortcuts to create cloud services such as ADW. The free cloud services are labeled Always Free Eligible. The Start Exploring section lists useful Oracle cloud online documentation. To initiate any cloud service, you select it from the pull-down menu under the Oracle Cloud logo. The free trial account can be upgraded to a paid account anytime. When the free trial is over, the account is limited to those Always Free resources. Once an account is upgraded to a paid account, your credit card is charged based on usage after the trial credits run out.

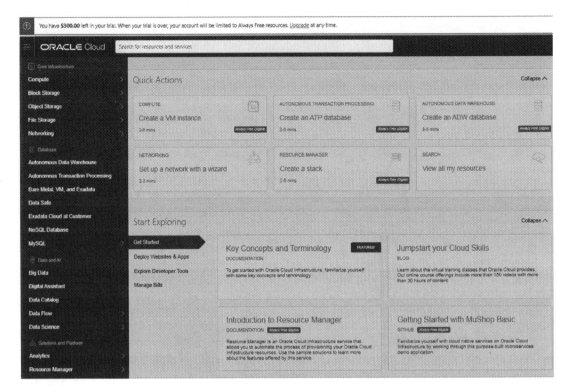

Figure 4-2. *Oracle Cloud Infrastructure page*

Oracle Autonomous Database Architecture and Components

This section provides an architecture overview of the main components of the Oracle Autonomous Database. Oracle Autonomous Database Cloud services provide fully autonomous database features, including ADW, Autonomous Transaction Processing, and Autonomous JSON Database. Features like self-driving, self-securing, and self-repairing fully automate provisioning, management, and maintenance of the database services. You can provision a database service with an easy-to-use graphical user interface (GUI) tool, load the data, and connect your applications to ADW.

In addition to the core database service, Oracle Autonomous Database provides a collaborative set of tools. For example, ADW includes data integration and business analytics tools, as shown in Figure 4-3.

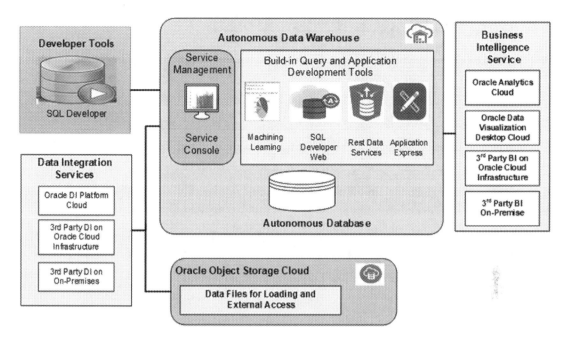

Figure 4-3. *ADW architecture*

ADW is optimized for data warehouse workloads. All the applications that work with Oracle Database should also work with ADW. These applications can be on-premise in your data center or on a public cloud. Figure 4-3 shows that development tools such as SQL Developer and Data Integrations Services connect to ADW for development or data integration. Also, business intelligence application services support connections with ADW. For example, by connecting to ADW, Oracle Analytics Cloud, or Oracle Analytics Desktop can create business data visualization and analytics.

Oracle Autonomous Warehouse includes a cloud-based service console and several cloud-based built-in query and applications tools. The service console is designed for managing tasks such as starting, stopping, and scaling the database service, and service monitoring tasks such as monitoring recent activities and database resource utilization. The built-in query and application development tools include the Oracle Machine Learning, SQL Developer Web version, Oracle REST Data Services, and Oracle Application Express. We discuss these tools further at the end of this chapter.

Oracle Autonomous Database Attributes

As a part of Oracle Cloud database services, Oracle Autonomous Database is a fully managed, preconfigured database environment. It is designed to bring full automation to the entire database life cycle. Three autonomous attributes—self-driving, self-securing and self-repairing—are embedded in the Autonomous Database services.

Self-driving attributes include rapid provisioning of the database services, self-scaling and automatic tuning, and automatic indexing. With these features, you don't need to configure any hardware infrastructure and install any software such as operating system (OS) and Oracle software or run database creation to provision Oracle Autonomous Database. All you need to do is to select the workload type and specify the sizing information, such as the number of CPU cores, the memory size, and the storage size of the database.

Oracle Autonomous Database service automates the database life-cycle management tasks, including the database creation, management, and maintenance tasks of the database such as database monitoring, tuning, and indexing, backup/recovery, patching, and upgrading. After the database is provisioned, you can scale the CPU core number or the memory size, or the storage size at any time.

Self-securing attributes ensure all the data in the Autonomous Database is stored in an encrypted format. The data access is only allowed by the authenticated users and applications through the certified-based authentication and Secure Sockets Layer (SSL). The certificate-based authentication uses an encrypted key stored in a wallet on both the client and the server. The critical security patches are automatically applied on-demand on different levels in the stack, including firmware, OS, hypervisor, Oracle Clusterware, and Oracle Database.

Self-repairing means self-healing hardware and software features. The self-healing hardware feature is based on the Exadata infrastructure's unique server failure detection, continuous monitoring of failing devices, and subsecond I/O storage redirection. The self-healing software relies on monitoring tools and fault prevention tools running in the background 24x7 and automatic problem detection and resolution. It applies known-problem diagnoses and resolutions with machine learning algorithms. Self-repairing also benefits from the maximal availability architecture, which protects Autonomous Database from unplanned downtime caused by hardware and software failure and planned downtime for hardware and software maintenance.

All these autonomous features automate the life-cycle management of the database and its underneath cloud infrastructure. This allows database users and application developers to focus on developing applications instead of managing and maintaining the hardware and software stacks.

Autonomous Database in Free Trier and Always Free

If you decide to use the Always Free option for your Oracle Cloud account, you can create and access up to two instances of either ADW or ATP. To ensure an Autonomous Database instance remains free, it must not exceed the following resource and capacity limitations.

- **Processor**: 1 Oracle CPU processor (cannot be scaled)

- **Memory**: 8 GB RAM

- **Database Storage**: 20 GB storage (cannot be scaled)

- **Workload Type**: The choice of either the transaction processing or data warehouse workload type

- **Database Version**: Oracle Database 21c or Oracle Database 19c

- **Infrastructure Type**: Shared Exadata infrastructure

- **Maximum Simultaneous Database Sessions**: 20

If an Always Free Autonomous Database has no activity for seven days, the database is shut down automatically. It can be restarted anytime. However, if an "Always Free" Autonomous Database is in a down state for three consecutive months, the database is automatically terminated and removed, and the database's resources are reclaimed.

Working with Oracle Autonomous Data Warehouse

This section describes how to work with the collaborative environment of ADW to support business analytics applications and machine learning. The Autonomous Transaction Processing database works in the same way as ADW does.

Working with Oracle Autonomous Data Warehouse starts with creating an ADW instance. The following are some of the typical tasks.

- Provisioning ADW

- Connecting to and querying ADW

- Loading or exporting data from/to ADW

Provisioning Oracle Autonomous Data Warehouse

After logging in to OCI, you go to the main page (see Figure 4-4).

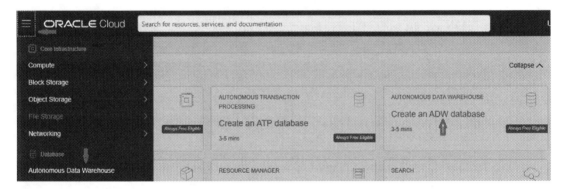

Figure 4-4. Oracle Cloud page

To create an ADW instance, go to the Oracle Cloud Infrastructure page and click "Create an ADW database" in Quick Action, or click Autonomous Data Warehouse in the navigation menu on the left side of the Oracle Cloud page. The arrows in Figure 4-4 highlight where these options are located.

On the Autonomous Database dialog page (see Figure 4-5), provide the following information for Autonomous Data Warehouse.

1. Enter the compartment, database name, and display name.

2. Select Data Warehouse in Workload Type.

3. Select Deployment type: Shared Infrastructure or Dedicated Infrastructure.

4. Determine whether to show only Always Free configuration options.

5.   Choose the database version. You can choose either 19c or 21c.

6.   Specify the OCPU count and the storage size in TB. If you select
the Always Free option, it takes the default fixed configuration:
OCPU=1 and Storage=0.02TB.

7.   If Auto Scaling is selected, it is enabled.

8.   Create administrator credentials. Enter the database admin
account password.

9.   Choose network access. Allow secure access from everywhere
unless you want to configure private access only using a virtual
cloud network.

10.   Choose a license type. Select Bring Your Own License or License
Included.

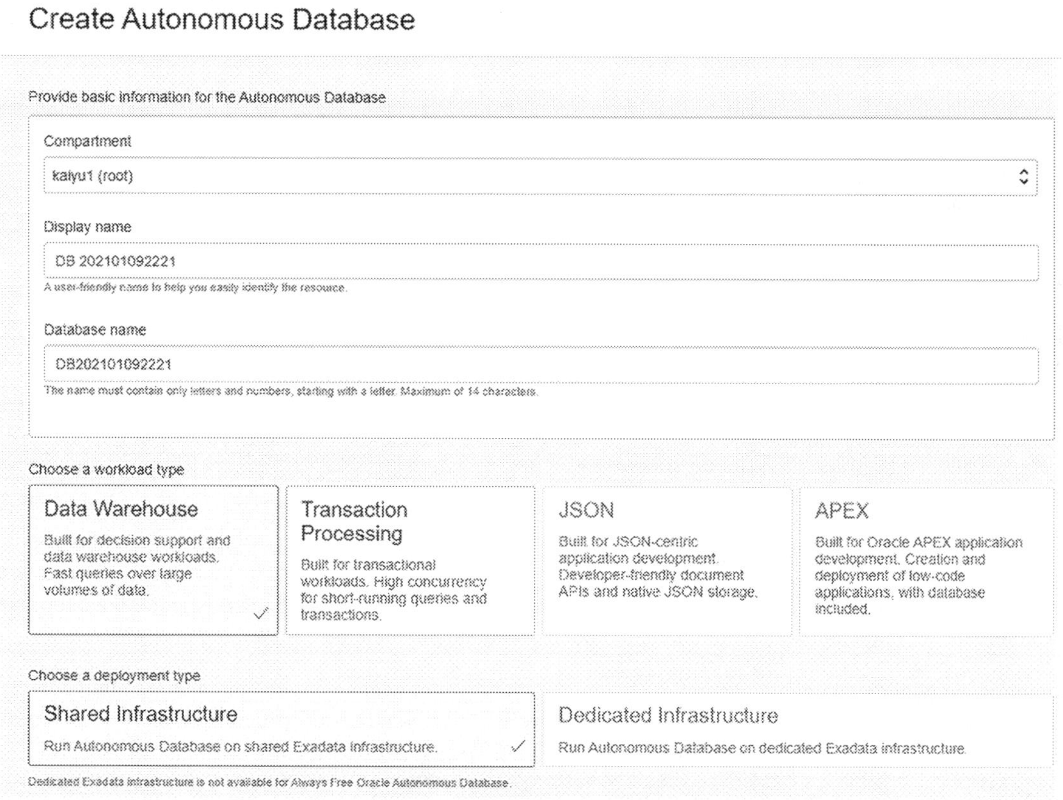

Figure 4-5. *Autonomous Database dialog page*

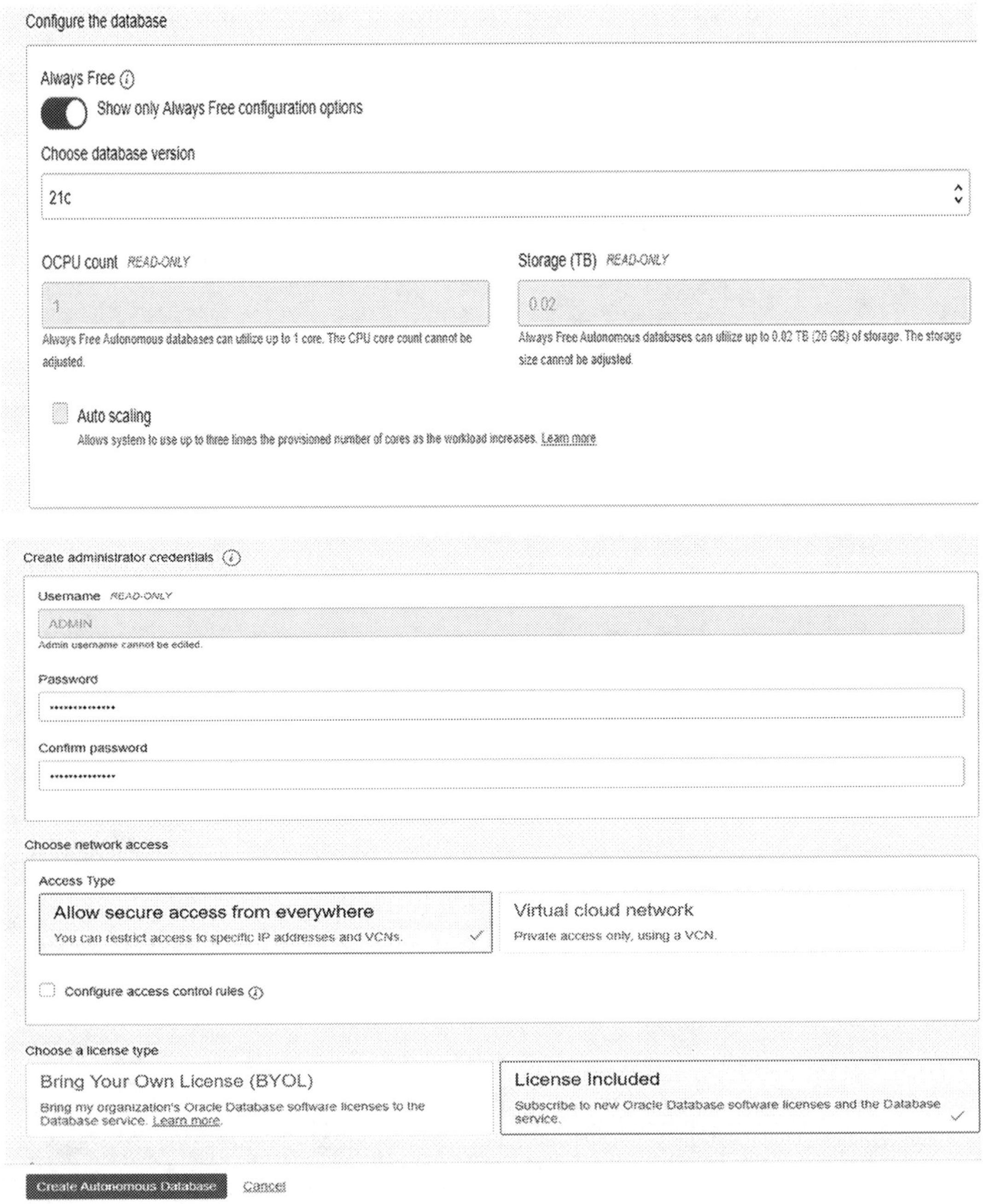

Configure the database

Always Free ⓘ

⬤⚪ Show only Always Free configuration options

Choose database version

21c

OCPU count *READ-ONLY*

1

Always Free Autonomous databases can utilize up to 1 core. The CPU core count cannot be adjusted.

Storage (TB) *READ-ONLY*

0.02

Always Free Autonomous databases can utilize up to 0.02 TB (20 GB) of storage. The storage size cannot be adjusted.

☐ Auto scaling
   Allows system to use up to three times the provisioned number of cores as the workload increases. Learn more

Create administrator credentials ⓘ

Username *READ-ONLY*

ADMIN

Admin username cannot be edited.

Password

•••••••••••••

Confirm password

•••••••••••••

Choose network access

Access Type

Allow secure access from everywhere   Virtual cloud network
You can restrict access to specific IP addresses and VCNs. ✓   Private access only, using a VCN.

☐ Configure access control rules ⓘ

Choose a license type

Bring Your Own License (BYOL)   **License Included**
Bring my organization's Oracle Database software licenses to the   Subscribe to new Oracle Database software licenses and the Database
Database service. Learn more.   service.   ✓

[Create Autonomous Database]   Cancel

Figure 4-5. *(continued)*

Click Create Autonomous Database. The database instance is automatically provisioned in about 5 to 10 minutes, depending on the size of the storage.

Connect to Oracle Autonomous Data Warehouse

Once an ADW database is successfully created, you see the word AVAILABLE under the ADW logo (see Figure 4-6). You see *Always Free* next to the database instance name, which indicates that the database is always available and free.

When establishing a connection to an ADW database over the public Internet, an authentication certificate, and Secure Sockets Layer (SSL) ensure the communication between the database client and database server is fully encrypted and secure. The certificate authentication uses an encrypted key stored in a wallet on both the client side and the database server side, which is your ADW instance. The key stored on the client side must match the key stored in the database server before the communication can be established. To configure this secure connection to the ADW instance, you need to download the wallet, which is a credential zip file, by clicking the DB Connection button, as shown in Figure 4-6.

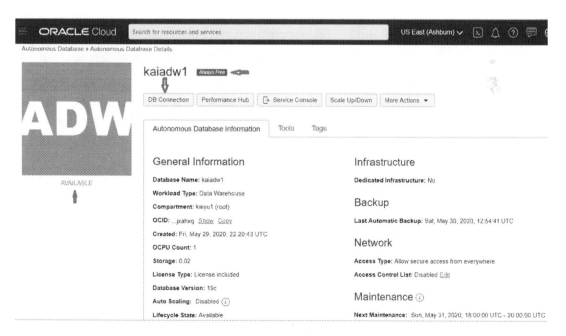

Figure 4-6. *Autonomous Database Information page*

This leads to the Database Connection page, as shown in Figure 4-7. You can click the Download Wallet to download the wallet zip file to your local computer. As shown in Figure 4-8, the wallet zip file (wallet_databasename.zip) includes the following:

- `tnsnames.ora` and `sqlnet.ora`: Network configuration files storing connect descriptors and SQL*Net client-side configuration.

- `cwallet.ora` and `ewallet.p12`: Auto-open SSO wallet and PKCS12 file. The PKCS12 file is protected by the wallet password provided in the UI.

- `keystore.jks` and `truststore.jks`: JKS Truststore and Keystore that is protected by the wallet passport provided while downloading the wallet.

- `ojdbc.properties`: Contains the wallet-related connection property required for JDBC connection. This should be in the same path as `tnsnames.ora`.

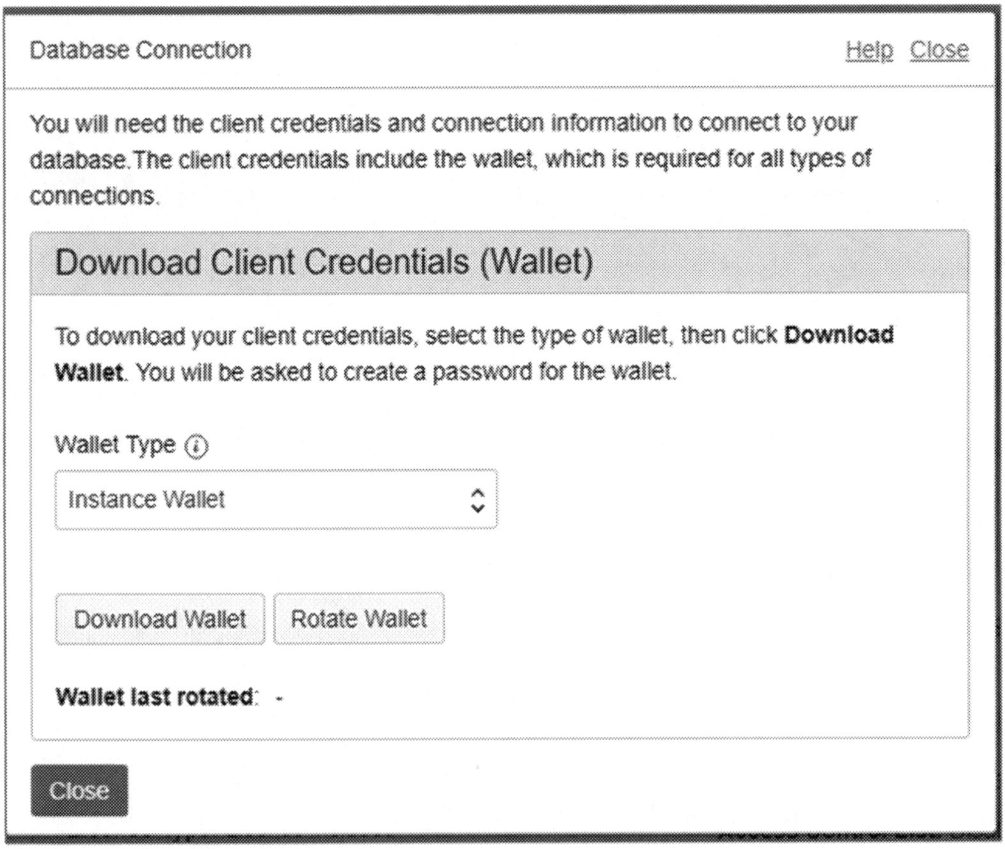

Figure 4-7. Download wallet

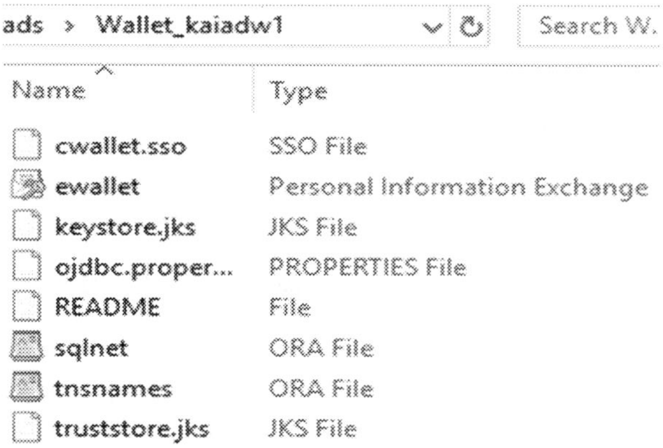

Figure 4-8. *Zipped wallet files*

To connect a client to the database, the wallet zip file needs to be included as part of the connection configuration. For example, if you want to connect to the database with an Oracle client, such as Oracle SQL*Plus, follow these steps.

1. Unzip the wallet_databasename.zip file to a directory; for example, /home/oracle/adw.

2. cd $ORACLE_HOME/network/admin, add the two entries wallet_location and SSL_SERVER_DN_MATCH to the sqlnet.ora file as shown in Listing 4-1. The wallet_location *entry* specifies the physical file location of the Oracle wallet, which includes certificates, keys, and trustpoints processed by SSL. The Setting of SSL_SERVER_DN_MATCH=Yes enforces that the distinguished name (DN) for the database server matches its service name.

Listing 4-1. Two Entries in sqlnet.ora File

```
WALLET_LOCATION = (SOURCE = (METHOD = file) (METHOD_DATA =
(DIRECTORY="/home/oracle/adw")))
SSL_SERVER_DN_MATCH=yes
```

3. Append the database service entries in /home/oracle/adw/ tnsnames.ora to $ORACLE_HOME/network/admin/tnsnames.ora.

   For example, append this database service name entry as shown in Listing 4-2 to $ORACLE_HOME/network/admin/tnsnames.ora.

Listing 4-2. Database Service Name Entry for ADW

```
kaiadw1_low = (description= (retry_count=20)(retry_delay=3)
(address=(protocol=tcps)(port=1522)(host=adb.us-ashburn-1.oraclecloud.com))
(connect_data=(service_name=mojg gjecekbigkm_kaiadw1_low.adwc.oraclecloud.com))
(security=(ssl_server_cert_dn="CN=adwc.uscom-east-1.oraclecloud.com,OU=Oracle
BMCS US,O=Oracle Corporation,L=Redwood City,ST=California,C=US")))
```

Usually the tnsnames.ora file from wallet zip file contains three database service name entries marked as high, medium, and low: <database_name>_low, <database_name>_medium and <database_name>_high. These predefined service names represent different levels of resource and concurrency that can be obtained for the database connection. You can use one of these service names to connect to the ADW database based on the performance and concurrency requirements for the database connection. For example, use this kaiadw1_low database service name for a database connection to the kaiadw1 ADW database with the lowest level of resources to each SQL statement and the greatest number of concurrent SQL statements. Listing 4-3 shows the SQL*Plus connection using admin user and this kaiadw1_low database service name.

Listing 4-3. SQL*Plus Connection to an ADW Instance

```
$sqlplus admin/admin_passwd@kaiadw1_low
```

To connect a client application with support for wallets (e.g., SQL Developer 18.3 or higher), you can select Cloud Wallet as the connection type and provide the wallet zip file in the configuration file box, as shown in Figure 4-9.

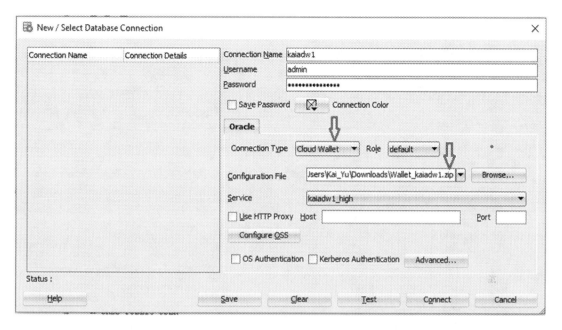

Figure 4-9. *Connect SQL Developer with ADW*

Once Oracle SQL Developer is connected to the ADW instance, you can use it to execute the SQL statements or browse the database schema objects and do many other tasks on the ADW, as shown in Figure 4-10.

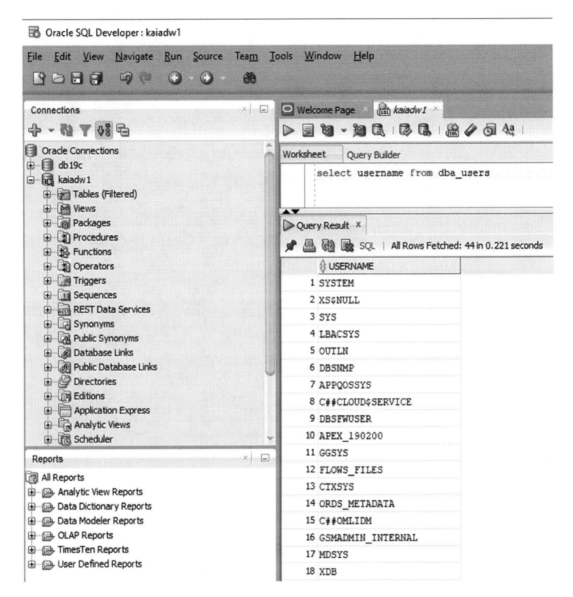

Figure 4-10. *Execute SQL query on ADW through Oracle SQL Developer*

Loading Data to Oracle Autonomous Data Warehouse

To start application development or machine learning projects with your ADW database, you may need to load data to the database from various data sources, such as loading a text file in .csv format to the data warehouse. In the Oracle Autonomous Data Warehouse environment, this data loading process consists of three steps, as shown in Figure 4-11.

1. Upload a data file from a client's local computer to Object Storage.

2. Create a proper cloud credential in ADW.

3. Copy the data file from Object Storage to a table in ADW.

Let's discuss the details of each step in the following sections.

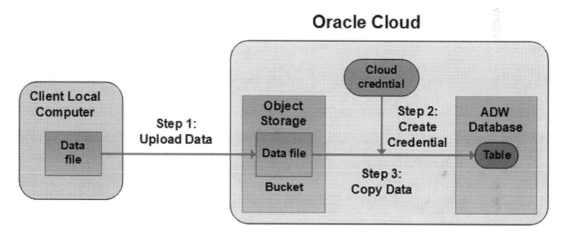

Figure 4-11. *Three steps to load data to Oracle ADW*

Step 1: Upload a File from a Local Computer to Object Storage

Object Storage uses buckets to organize files. To load a file to Object Storage, you must first create a bucket inside it. Perform the following steps in the Oracle Cloud console to create a bucket.

1. On the Oracle Cloud main page, navigate to Object Storage and click Create Bucket (see Figure 4-12).

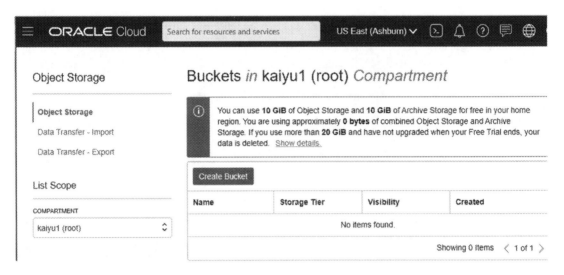

Figure 4-12. *Object Storage: Create bucket*

2. In the Create Bucket dialog box, enter the bucket name (e.g., *data1*). Select the storage tier, and click Create Bucket. Figure 4-13 shows that the data1 bucket is created.

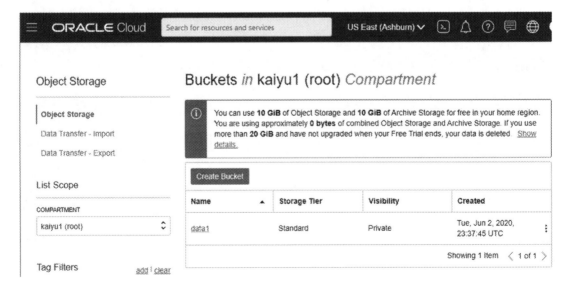

Figure 4-13. *Bucket data1 created*

Do the following to upload a file to a bucket in the Buckets window, as shown in Figure 4-13.

1. Click the bucket name (*data1*) to get the bucket's details, and then click Upload Objects (see Figure 4-14).

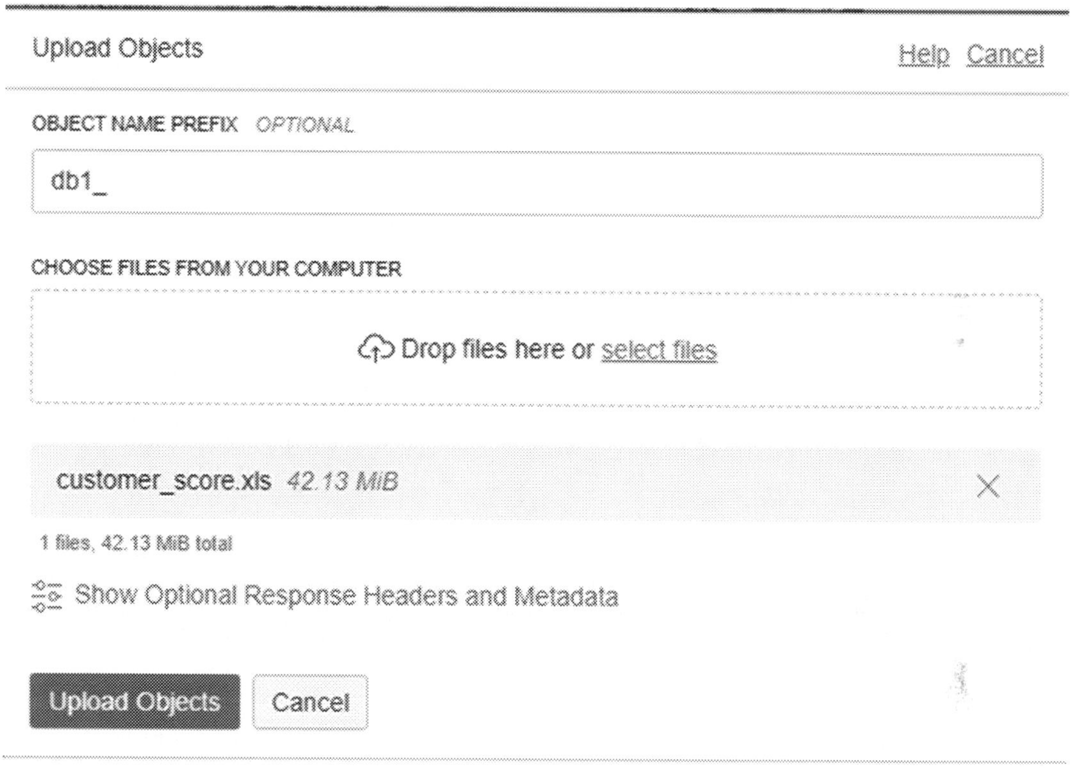

Upload Objects   Help Cancel

OBJECT NAME PREFIX *OPTIONAL*

db1_

CHOOSE FILES FROM YOUR COMPUTER

⌂ Drop files here or select files

customer_score.xls *42.13 MiB*   ✕

1 files, 42.13 MiB total

⚏ Show Optional Response Headers and Metadata

Upload Objects   Cancel

Figure 4-14. *Upload objects*

2. Drag a file into the "Drop files here … or select files" dialog box.

3. Optionally, specify the file name prefix in the object name prefix field.

4. Click Upload Objects to load the file to the Object Storage bucket.

5. After uploading the file, get the object details of this file by clicking View Object Details, as shown in Figure 4-15. The object details are shown in Listing 4-4 shows the object details.

Objects

Figure 4-15. *View Object Details*

Listing 4-4. Object Details

```
Object Details
Basic Information
Name: db1_customers.csv
URL Path (URI): https://objectstorage.us-ashburn-1.oraclecloud.com/n/
idudzqoitv0b/b/data1/o/db1_customers.csv
Storage Tier: Standard
Size: 2.14 MiB
```

One of the most important details is the URL Path (URI) of the object. Keep the URL path because it is used in the DBMS_CLOUD.COPY_DATA procedure.

Step 2: Create a Credential

To load the data from a file stored in Object Storage to an ADW, you need to create an Object Storage credential in ADW. Use your OCI login name and an auth token.

After you sign up for an Oracle Cloud account and log in to Oracle Cloud, perform the following steps to create an auth token for your OCI login.

1. On the Oracle Cloud main page, open the Profile menu, ![profile icon], located in the top-right corner. Then click User Settings (see Figure 4-16) to go to the User Details page.

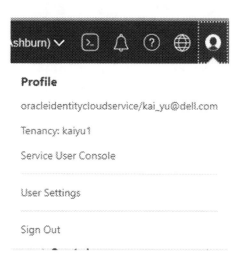

Figure 4-16. *Profile menu*

2. On the User Details page (see Figure 4-17), look for the Resources section and click Auth Tokens.

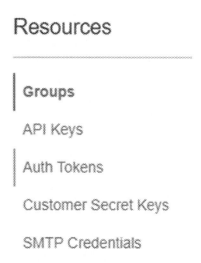

Figure 4-17. *User Details page: Auth Tokens*

3.   In the Auth Tokens section, click Generate Token (see Figure 4-18).

Figure 4-18.   *Generate token*

4.   In the Generate Token box, enter a description of the token; for
     example, ADW_TOKEN.

5.   Click the Generate Token button. The new token string is
     displayed (see Figure 4-19).

Figure 4-19.   *Copy token string*

6.   Copy the 'nL7HEjMP{3>:;n40VSdS' token string immediately.
     With the ADW_TOKEN token name and the token string, create
     an Object Storage credential with the DBMS_CLOUD.CREATE_
     CREDENTIAL procedure.

In the CREATE_CREDENTIAL procedure (see Listing 4-5), the password parameter
takes the 'nL7HEjMP{3>:;n40VSdS' token string. The username is the OCI account login
name.

Listing 4-5. Store Object Storage Credentials in ADW

```
BEGIN
DBMS_CLOUD.CREATE_CREDENTIAL(credential_name => 'ADW_TOKEN',
 username => 'oracleidentitycloudservice/kai_yu@dell.com',
 password => 'nL7HEjMP{3>:;n4OVSdS');
END;
/
```

Step 3: Load Data to a Table in Autonomous Data Warehouse

After loading the data into the Oracle Cloud's Object Storage service, you can load this data to the ADW database. First, you need to create a customer table by executing the DDL in Listing 4-6.

Listing 4-6. Create the Customer Table to Store the Data

```
create table customer
    (
    customer number(38,0),
    age number(4,0),
    income number(38,0),
    marital status varchar2(26 byte)
    );
```

Second as shown in Listing 4-7, Execute the DBMS_CLOUD.COPY_DATA procedure to copy the data from the db1_customer_score.xls file in the Object Storage to the customer table in the ADW. This DBMS_CLOUD.COPY_DATA procedure takes several parameters.

- table_name: The table where the data is copied to

- credential_name: The credential name is created with the DBMS_CLOUD.CREATE_CREDENTIAL

- file_uri_list: The URL address of the source file stored in the bucket of the Object Storage obtained from the object's Object Details, as shown in Listing 4-4

- Format: Describes the format of the source files (options are specified as a JSON string)

121

Listing 4-7. Load the Data from Object Storage to Customer Table

```
BEGIN
DBMS_CLOUD.COPY_DATA (
    table_name =>'CUSTOMER',
    credential_name =>'ADM_TOKEN',
    file_uri_list =>'https://objectstorage.us-ashburn-1.oraclecloud.com/n/
    idudzqoitvOb/b/data1/o/db1_customers.csv',
    format =>json_object('ignore missing columns' value 'true', 'removequotes'
    value 'true', 'blankasnull' value 'true', 'delimiter' value ',')
);
END;
/
```

If you ever run into any issue with the COPY_DATA procedure, you can either review the error message from the execution of the procedure for diagnosing purposes. Listing 4-8 is an example of the error message.

Listing 4-8. DBMS_CLOUD.COPY_DATA Procedure Error Message

```
Error report -
ORA-20003: Reject limit reached, query table "ADMIN"."COPY$11_LOG" for
error details
ORA-06512: at "C##CLOUD$SERVICE.DBMS_CLOUD", line 943
```

This error message indicates that you can find more details of the error message by querying the error log table COPY$11_LOG table. Listing 4-9 shows a query on the error log table: ***COPY$11_LOG to get more detailed messages.***

Listing 4-9. Query Error Log Table COPY$11_LOG

```
Select * from COPY$11_LOG;
ORA-29913: error in executing ODCIEXTTABLEOPEN callout
ORA-29400: data cartridge error
```

This message indicates some data error. After further investigating the data, the data exception is due to some blank fields in the data. The solution was to add the 'blankasnull' value 'true' option to the JSON object of the format parameter of DBMS_CLOUD.COPY_DATA, as shown in Listing 4-7, which we initially did not use.

Import Tables/Schema to Oracle Autonomous Database

It is common to migrate a schema or tables from a local on-premise database to ADW. A quick way to do this type of migration is to export the schemas or tables from the local database to dump files and then import the dump files to ADW. For a fast and stable import, it is recommended that the dump files be uploaded to Oracle Cloud Object Storage and then imported into the ADW database instance. Figure 4-20 illustrates the four steps of this migration process.

1. Export the schema or the tables from a source database to dump files.

2. Upload the dump files to the Object Storage in Oracle Cloud.

3. Create a Cloud Authentication Credential.

4. Import the schema or tables from dump files to ADW. The import process takes place on a client computer.

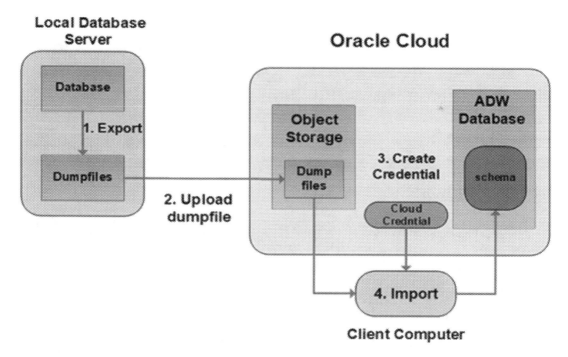

Figure 4-20. *Export and import schema/tables to ADW*

The following are the four steps in detail.

Step 1. Export from a Source Database

Oracle ADW document [1] suggests that two export parameters—EXCLUDE and DATA_OPTIONS—be set for data dump export. Listing 4-10 shows the settings for these two parameters.

Listing 4-10. Setting Two Parameters for Data Dump Export

```
exclude=index,cluster,indextype,materialized_view,materialized_view_
log,materialized_zonemap,db_link
encryption_pwd_prompt=yes
```

These settings ensure that those object types that are not required in ADW are excluded in the export and ensure that table partitions are grouped together for faster import to ADW. Listing 4-11 is an example of exporting sales schema from database K19cdb.

Listing 4-11. Export Sales Schema from the Source Database

```
$expdp sales/passwd@K19cdb  exclude=index,cluster,indextype,materialized_
view,materialized_view_log,materialized_zonemap,db_link \
data_options=group_partition_table_data \
parallel =2 \
schemas=sales \
dumpfile=export%u.dmp \
encryption_pwd_prompt=yes \
log=expdp_k19c.log
```

Here, since the parallel parameter is set to 2, only two dump files, export01.dmp and export02.dmp, are generated. If the ADW instance has more CPUs, the parallel parameter can be set to that number. For example, if there are 16 CPUs, the export generates 16 dump files which allow you to import them to the ADW instance in 16 parallel streams for a faster import.

To increase the security for the database export, you can set the encryption_pwd_prompt parameter as Yes. With this setting, the export prompts to enter a password for encrypting the dump files.

When importing the dump files, you are prompted to provide the same encryption password for decrypting the dump files.

Step 2. Upload the export01.dmp and export02.dmp dump files to Object Storage. The method to do this is the same as the one covered in step 1.

Step 3. Create a cloud authentication credential. Let's use the same method used in step 2 and Listing 4-5 of the last section to create the ADW_TOKEN credential.

Step 4: Import to Autonomous Data Warehouse.

As shown in Figure 4-20, the import operation occurs on a local computer with an Oracle client software installation. To connect to the Autonomous Datawarehouse database from the local computer where the import will take place, let's add two entries to $ORACLE_HOME/network/admin/sqlnt.ora, as shown in Listing 4-12.

Listing 4-12. Two Entries in sqlnet.ora File

```
WALLET_LOCATION = (SOURCE = (METHOD = file)
(METHOD_DATA = (DIRECTORY="/home/oracle/adw")))
SSL_SERVER_DN_MATCH=yes
```

The ADW database service name entry is copied to $ORACLE_HOME/network/admin/tnsnames.ora in the Oracle client installation. Listing 4-13 is an example of an ADW database service name entry.

Listing 4-13. ADW Database Service Name Entry

```
kaiadw1_low = (description= (retry_count=20)(retry_delay=3)
(address=(protocol=tcps)(port=1522)(host=adb.us-ashburn-1.oraclecloud.com))
(connect_data=(service_name=mojg gjecekbigkm_kaiadw1_low.adwc.oraclecloud.
com))(security=(ssl_server_cert_dn="CN=adwc.uscom-east-1.oraclecloud.
com,OU=Oracle BMCS US,O=Oracle Corporation,L=Redwood City,ST=California,
C=US")))
```

The SQLPLUS command can test the connection to the ADW instance, as shown in Listing 4-14.

Listing 4-14. SQLPLUS Login to ADW Instance

```
$sqlplus admin/admin_passwd@kaiadw1_low
```

Before importing the dump files to the ADW database, you need to ensure that the following prerequisites are met.

- Both export01.dmp and export02.dmp are uploaded into Object Storage in step 2.

- The URLs of the dump files (see Listing 4-15) are copied for import operation.

Listing 4-15. URLs for the Two Dump Files

```
https://objectstorage.us-ashburn-1.oraclecloud.com/n/idudzqoitvOb/b/
data1/o/export01.dmp
https://objectstorage.us-ashburn- 1.oraclecloud.com/n/idudzqoitvOb/b/
data1/o/export02.dmp
```

The ADW_TOKEN credential is created in step 3. With all these prerequisites being met, the `impdp` command (see Listing 4-16) imports the schema from the dump files in Object Storage to the ADW database.

Listing 4-16. Import to ADW

```
impdp admin/admin_passwd@kaiadw1_low \
directory=data_pump_dir \
credential=ADW_TOKEN \
dumpfile=https://objectstorage.us-ashburn-1.oraclecloud.com/n/
idudzqoitvOb/b/data1/o/export01.dmp, https://objectstorage.us-ashburn-1.
oraclecloud.com/n/idudzqoitvOb/b/data1/o/export02.dmp  \
parallel=2 \
encryption_pwd_prompt=yes \
partition_options=merge \
```

By default, the import log file import.log is stored in data_pump_dir, which was specified with an impdp directory parameter. You can check the directory path for data_pump_dir by running this query in the ADW using its admin account. For example, this query result is like the one in Listing 4-17.

Listing 4-17. data_pump_dir Directory Path

```
Select directory_path from all_directories where directory_name =
'DATA_PUMP_DIR';

DIRECTORT_PATH
-------------------------
/u03/dbfs/A6B99C1F8E42C676E0539114000A16AF/data/dpdump
```

However, since this file system of the ADW host is not directly accessible, instead of accessing the import.log file directly, you need to move this file to Object Storage, then download this file from Object Storage. Listing 4-18 shows how to use DBMS_CLOUD. PUT_OBJECT procedure to put the import.log file to Object Storage.

Listing 4-18. Put the Import Log to the Object Storage

```
BEGIN
DBMS_CLOUD.PUT_OBJECT(
    credential_name =>'ADW_TOKEN',
    object_uri =>'https://objectstorage.us-ashburn-1.oraclecloud.com/n/
    idudzqoitv0b/b/data1/o/import.log',
    directory_name   => 'DATA_PUMP_DIR',
    file_name        => 'import.log');
END;
/
```

After this procedure executes, the import.log is shown in Object Storage (see Figure 4-21). This file can be downloaded by clicking Download.

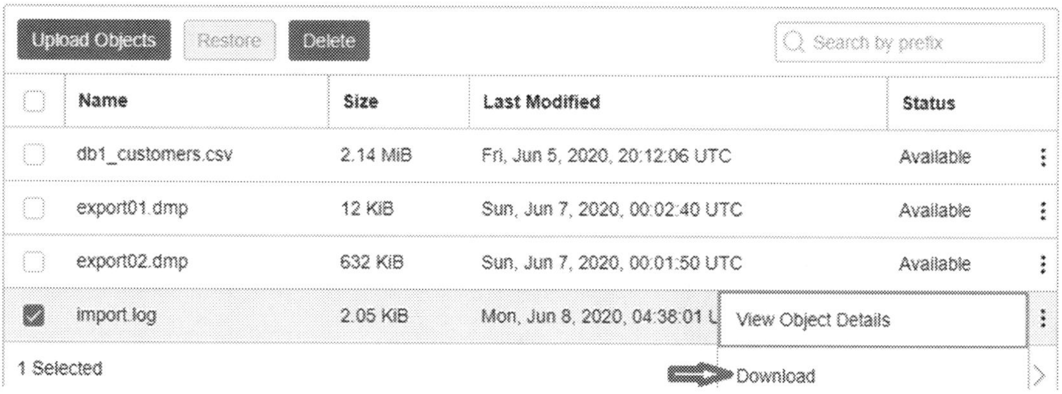

	Name	Size	Last Modified	Status	
☐	db1_customers.csv	2.14 MiB	Fri, Jun 5, 2020, 20:12:06 UTC	Available	⋮
☐	export01.dmp	12 KiB	Sun, Jun 7, 2020, 00:02:40 UTC	Available	⋮
☐	export02.dmp	632 KiB	Sun, Jun 7, 2020, 00:01:50 UTC	Available	⋮
☑	import.log	2.05 KiB	Mon, Jun 8, 2020, 04:38:01 U	View Object Details	⋮

1 Selected   Download

Figure 4-21. *Download Import.log from Object Storage*

Oracle Machine Learning with ADW

ADW includes the following built-in query and application development tools (also see Figure 4-3).

- **Oracle Machine Learning (OML)**: A web-based notebook application that provides simple querying, data visualization, and collaboration capabilities. The notebook can be used by developers, business users, and data scientists to perform data analytics, data discovery, and data virtualizations.

- **Oracle SQL Developer**: A web browser-based interface of Oracle SQL Developer.

- **Oracle REST Data Service (ORDS)**: A Java-based data service for developing modern REST interfaces for relational data and the JSON Document Store.

- **Oracle Application Express (APEX)**: A low-code development platform that enables you to build scalable, secure enterprise apps.

These built-in tools can be accessed through the Tools tab of the ADW Database Details page, as shown in Figure 4-22. We are focusing on the Oracle Machine Learning built-in tool. If you are interested in the other tools, refer to Oracle Cloud Using Oracle Autonomous Data Warehouse on Shared Exadata Infrastructure, F35573-24, June 2021, Oracle Corporation (URL: `https://docs.oracle.com/en/cloud/paas/autonomous-database/adbsa/`).

kaiadw1

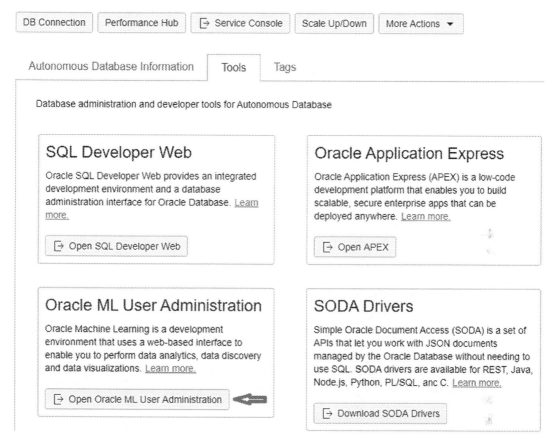

Figure 4-22. *Autonomous Database build tools*

Accessing Oracle Machine Learning Through Oracle Autonomous Database

Oracle Machine Learning is a built-in application for Oracle Autonomous Database. You can access it through an Oracle Autonomous Database such as an ADW instance by adding an existing database user account to Oracle Machine Learning or creating a new user account with Oracle Machine Learning management GUI.

Do the following to add a new user account to Oracle Machine Learning.

1. Click the Service Console in Autonomous Data Warehouse.

2. On the Service Console, click Administration.

3. In Administration, click Manager OML Users to open the Oracle Machine Learning User Administration page (see Figure 4-23) and then click Create to open the Create User page.

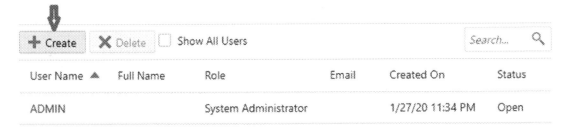

Figure 4-23. User administration page

4. On the Create User page, enter the required information, such as username, first and last name, email address, and password, and then click Create. A new user, OMLUSER, is created.

5. Click the Home logo on the top-right of the screen (see Figure 4-24). This takes you to the Oracle Machine Learning login page (see Figure 4-25), where you can log in to Oracle Machine Learning using the username and password you created.

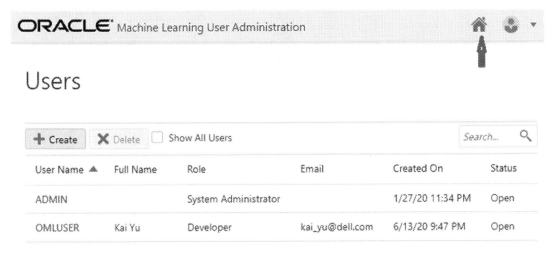

Figure 4-24. New user is created

ORACLE Cloud Infrastructure

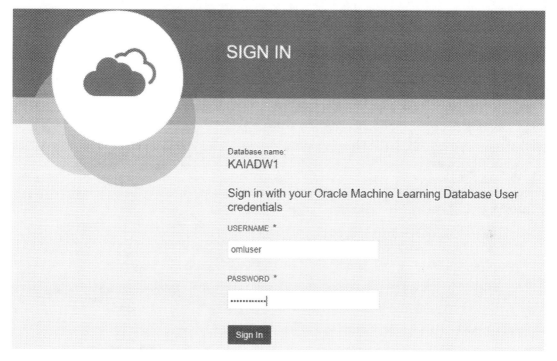

Figure 4-25. *Oracle Machine Learning login page*

You also can add an existing database user account to Oracle Machine Learning.

6. Click the service console in an Autonomous Database instance.

7. In the service console, click Administration.

8. In Administration, click Manager OML Users to open the Oracle Machine Learning User Administration page (see Figure 4-26), and then click Show All Users to open the Create User page.

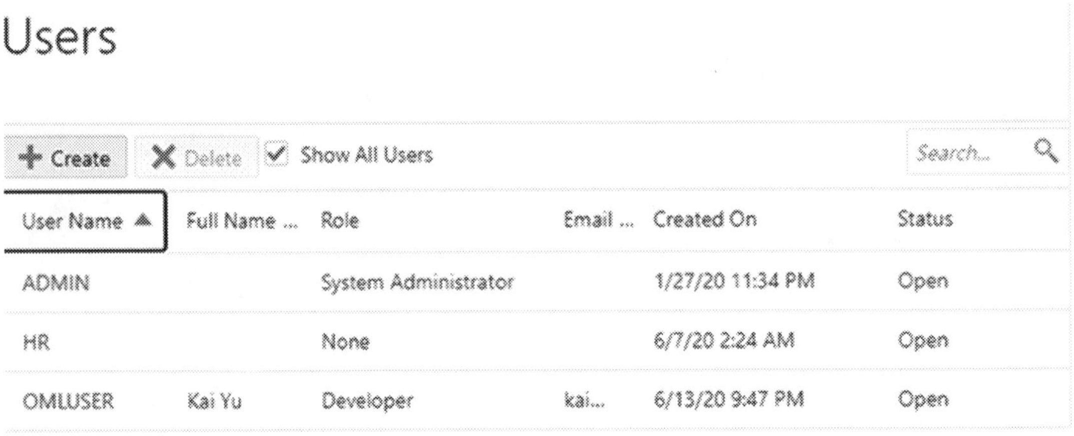

Users

User Name ▲	Full Name ...	Role	Email ...	Created On	Status
ADMIN		System Administrator		1/27/20 11:34 PM	Open
HR		None		6/7/20 2:24 AM	Open
OMLUSER	Kai Yu	Developer	kai...	6/13/20 9:47 PM	Open

Controls above table: + Create   ✕ Delete   ☑ Show All Users   Search... 🔍

Figure 4-26. *User administration page*

9. Select a user such as HR to click and go to the Edit User window (see Figure 4-27). Edit any field in the window and click Save. This results in granting the OML Developer role to the HR user, as shown in Figure 4-28.

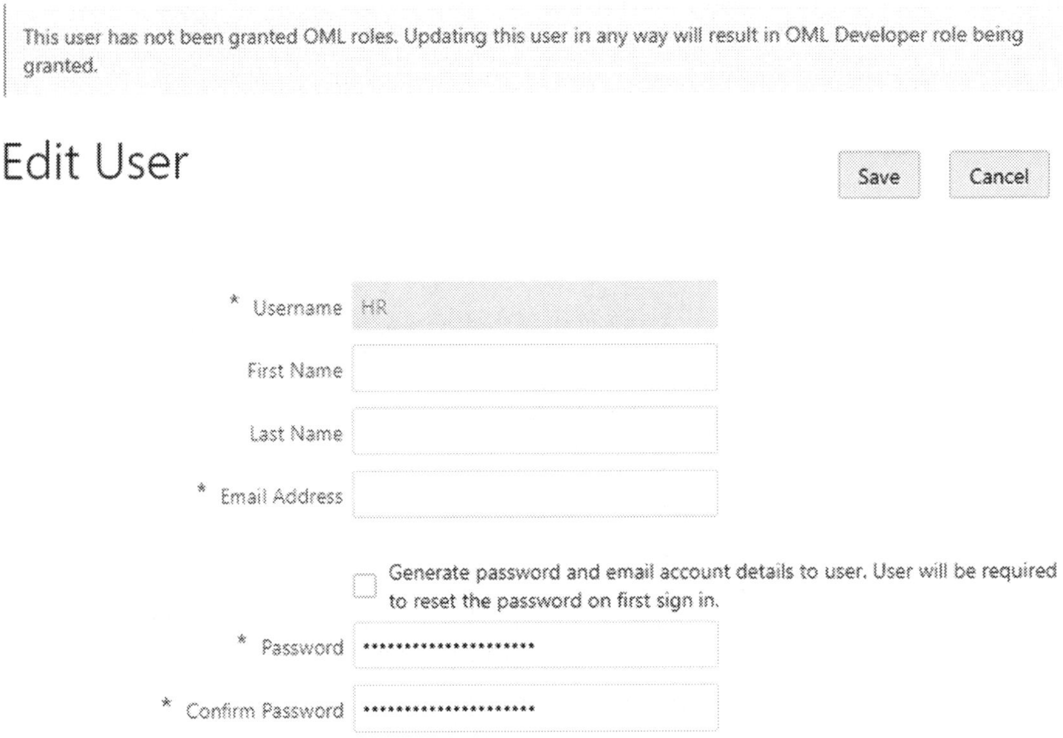

This user has not been granted OML roles. Updating this user in any way will result in OML Developer role being granted.

Edit User

Save   Cancel

* Username   HR

First Name   _____

Last Name   _____

* Email Address _____

☐ Generate password and email account details to user. User will be required to reset the password on first sign in.

* Password   ••••••••••••••••••••

* Confirm Password ••••••••••••••••••••

Figure 4-27. *Edit the existing user to grant the developer role*

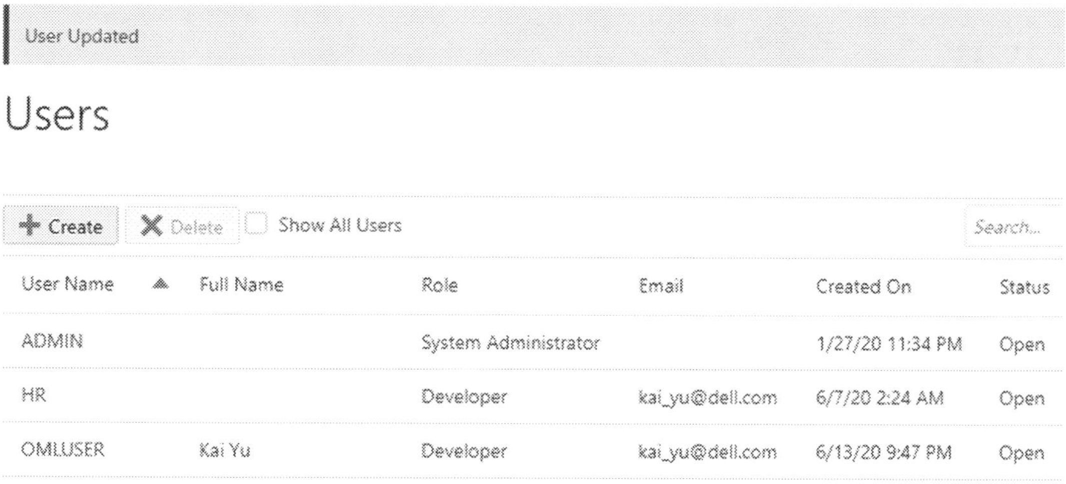

User Updated

Users

User Name	▲	Full Name	Role	Email	Created On	Status
ADMIN			System Administrator		1/27/20 11:34 PM	Open
HR			Developer	kai_yu@dell.com	6/7/20 2:24 AM	Open
OMLUSER		Kai Yu	Developer	kai_yu@dell.com	6/13/20 9:47 PM	Open

+ Create   ✕ Delete   ☐ Show All Users   Search...

Figure 4-28. *Existing user HR granted the OML Developer role*

Summary

This chapter introduced Oracle Cloud Infrastructure services with a special focus on Autonomous Database cloud service. We discussed how to sign up for a free tier account on OCI services and Autonomous Database such as ADW and how to achieve Always Free service to leverage these cloud services for ongoing machine learning projects. We also discussed how to perform some important tasks such as establishing database connection, loading data, and importing schema and tables to ADW, which can be considered the database skills essential to your machine learning project with ADW. At the end of the chapter, we introduced Oracle Machine Learning built-in tool in ADW. The next chapter focuses on developing machine learning projects with this tool.

CHAPTER 5

Running Oracle Machine Learning with Autonomous Database

The last chapter explored Oracle Machine Learning (OML), a built-in environment offered with Autonomous Database. We discussed how to add a database user account or grant an existing database user account with the machine learning developer role. This role enables a database account user to use the OML environment and Autonomous Database for a machine learning project. Oracle Machine Learning works with both types of Oracle Autonomous Databases: Autonomous Data Warehouse (ADW) and Autonomous Transaction Processing (ATP).

This chapter dives into the Oracle Machine Learning environment. It discusses the elements of the environment (workspace, project, and notebook) and how this environment can be shared between developers for collaborative OML development projects. We explore two important notebook settings: the database connection setting and the interpreter binding setting. The database connection setting specifies a notebook's link and its back-end database, in this case, Oracle Autonomous Database. Although the rest of the chapter focuses on ADW, the content also applies to ATP.

A notebook's interpreter bindings determine how the paragraphs are interpreted and executed. An important interpreter binding is with Oracle Database's SQL and PL/SQL interpreters, essentially the Autonomous Database. With the settings of interpreter binding and the database connection with the ADW database, you can create and run SQL statements and PL/SQL scripts in a notebook against the ADW database

© Heli Helskyaho, Jean Yu, Kai Yu 2021
H. Helskyaho et al., *Machine Learning for Oracle Database Professionals*,
https://doi.org/10.1007/978-1-4842-7032-5_5

on the back end. By leveraging the SQL and PL/SQL support in OML, the notebook, developers, data scientists, developers, and business analysts can accomplish some of the critical tasks of their machine learning project.

- Running PL/SQL scripts in a notebook allows Oracle Machine Learning for SQL (OML4SQL) packages to perform essential machine learning tasks, such as data transformation, model building, and model evaluation.

- Executing a SQL statement in a notebook fetches result data from the autonomous database. The resulting data can be displayed in various formats for data analytics, data discovery, and data visualization.

The following topics are covered in this chapter.

- OML collaborative environment

- Running SQL scripts and SQL statement

- Notebooks and data visualization

Oracle Machine Learning Collaborative Environment

Oracle Machine Learning Notebooks in Oracle Autonomous Databases like ADW provide a collaborative web-based interface for data analysis, data discovery, data visualization, and collaboration. This notebook environment is based on the Apache Zeppelin notebook, allowing data scientists and developers to collaborate to build, evaluate, and deploy machine learning models. You can write code and scripts and text in a notebook and execute them in an ADW database. Notebooks allow you to create data visualization and perform the data analytics from executed code or scripts, such as SQL query results or SQL scripts from ADW.

Starting with Oracle Machine Learning

To use Oracle Machine Learning in ADW, you need to log in to OML with an ADW database user account with the OML developer role. The admin user of the ADW database can create a new database user account with the OML developer role or simply grant the OML developer role to an existing user in ADW, as discussed in Chapter 3.

Once you log in to OML, you go to the home page (see Figure 5-1).

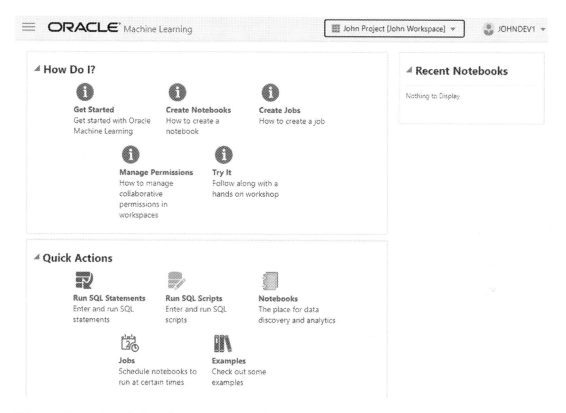

Figure 5-1. *Oracle Machine Learning home page*

There are two sections on the page. The How Do I? section provides links to many online documents and hands-on instructional materials. The Quick Actions section has links to the action pages. Clicking Notebooks, or Run SQL statements, or Run SQL Scripts leads to the notebook creation page, or the SQL Query Scratchpad page, or the SQL Script Scratchpad page, respectively. The Examples link lists many good examples that you can try to get hands-on experience with machine learning tasks. The Jobs link allows you to schedule jobs to run in Notebooks.

In the upper right of the home page, a text box, ▦ John Project [John Workspace] ▾ , shows the current project name and the workspace name. OML provides three levels of hierarchy to organize contents: workspace, project, and notebook. A workspace is an area where you can store your projects. A project can have one or many notebooks. In other words, a workspace is a container of projects, and a project is a container of notebooks.

When a user logs in to OML for the first time, an initial workspace and a default project are created. As shown in Figure 5-2, the new user login, Johndev1, is provided with the initial workspace, John Workspace, and the default project, John Project.

Name	Created By	Type	Created On	Updated By	Last Update
☑ John Workspace	JOHNDEV1	Work...	07/27/20 09:34:26 PM	JOHNDEV1	07/27/20 09:
John Project	JOHNDEV1	Project	07/27/20 09:34:26 PM	JOHNDEV1	07/27/20 09:

Figure 5-2. *Initial workspace and default project*

A user can also create additional workspaces and projects and select a different workspace and project to create a notebook. As shown in Figure 5-3, by clicking the upper-right textbox on the Oracle Machine Learning home page, you can see a project workspace drop-down list from which you can select a project or create a new project or manage the workspace and permission. The top of the list indicates the current project and workspace.

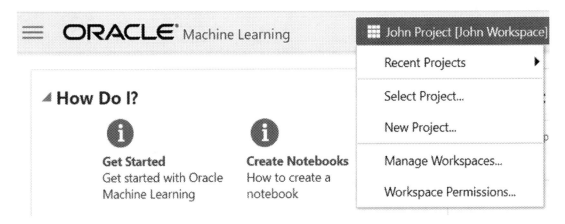

Figure 5-3. *Project workspace drop-down list*

To create a new project, you can either use the existing workspace or create a new workspace by clicking the + (plus sign), as shown in Figure 5-4.

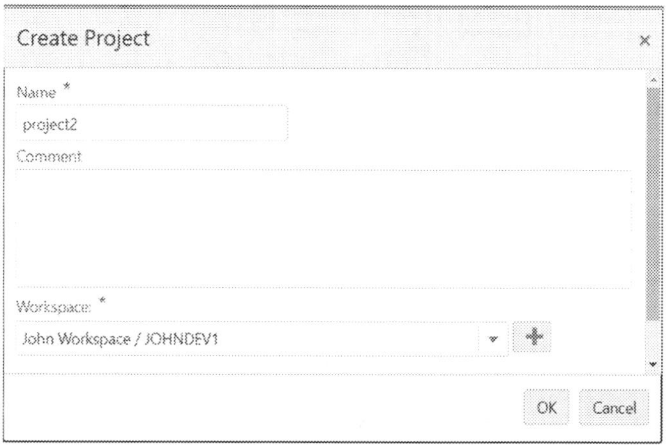

Create Project   ×

Name *

project2

Comment

Workspace *

John Workspace / JOHNDEV1   ▾   +

OK   Cancel

Figure 5-4. *Create project*

Figure 5-5 shows an example of two login users Johndev1 and Nancydev2, and their workspaces, projects, and notebooks in OML and how these are connected to the ADW database.

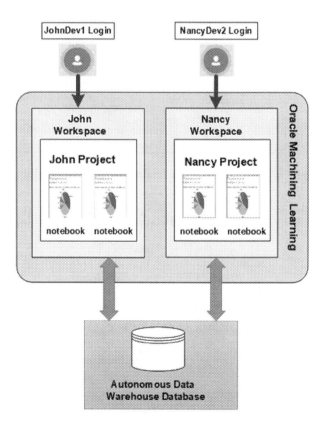

Figure 5-5. *User logins and their workspaces, projects, and notebooks*

Sharing Workspaces with Other Users

You can collaborate with other users by permitting them to access your workspace, projects, and notebooks inside of the workspace. You can grant three different types of workspace access permissions to other users.

- **Manager**: View-only for a workspace; create, update, and delete for a project; create, update, run, delete, and schedule job for a notebook. Be aware that a user with manager permission can also drop tables and run any scripts on the owner's account.

- **Developer**: View-only for a workspace; view-only for a project; create, update, run, or delete a notebook that the developer created. Be aware that a user with developer permission can also drop tables and run scripts on the owner's account. You can view and run jobs from shared notebooks only. A developer cannot create jobs for the notebooks shared.

To grant access permission to another user, click the Project and Workspace box on the upper-right corner of the OML home page to get a drop-down list. Then, select Workspace Permissions, as shown in Figure 5-6.

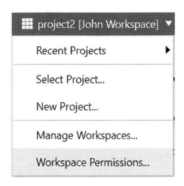

Figure 5-6. *Drop-down list*

In the Workspace Permissions window, select the user and permission type from a menu: Manager, Developer, and Viewer, as shown in Figure 5-7. Click +Add and OK.

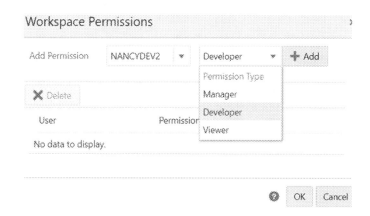

Figure 5-7. *Workspace permission window*

Next, the user receives permission to access the shared workspace. As shown in Figure 5-9, in addition to the default Nancy Workspace, the Nancydev2 user has access permission in John Workspace and two projects, John Project and project2.

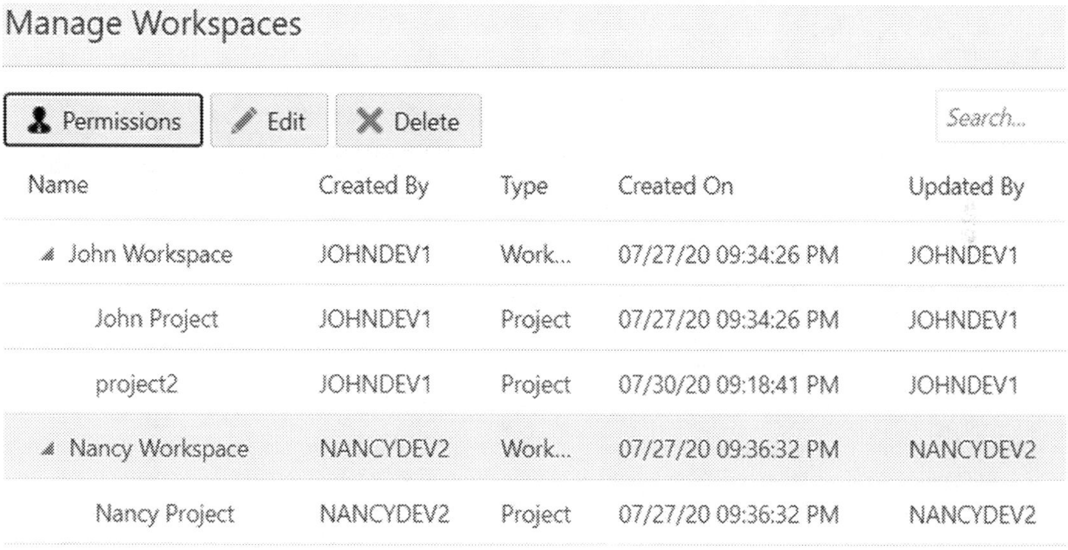

Manage Workspaces

Name	Created By	Type	Created On	Updated By
⊿ John Workspace	JOHNDEV1	Work...	07/27/20 09:34:26 PM	JOHNDEV1
John Project	JOHNDEV1	Project	07/27/20 09:34:26 PM	JOHNDEV1
project2	JOHNDEV1	Project	07/30/20 09:18:41 PM	JOHNDEV1
⊿ Nancy Workspace	NANCYDEV2	Work...	07/27/20 09:36:32 PM	NANCYDEV2
Nancy Project	NANCYDEV2	Project	07/27/20 09:36:32 PM	NANCYDEV2

Figure 5-8. *Nancydev2 has the access permission in John Workspace*

Creating a Machine Learning Notebook

Once you select the project or create a new project, you can create a notebook inside the project that you select. You start creating a notebook by either clicking the Notebook icon in the Quick Action section or selecting Notebooks in the pull-down menu on the upper left of the OML home page. This leads to the Notebooks page, as shown in Figure 5-9.

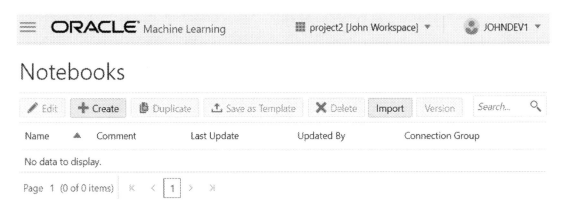

Figure 5-9. *Notebooks screen*

Clicking the Create button leads to the Create Notebook dialog box. You can specify the notebook name in the Name field and enter any comments in the Comments field. In the Connection field, you can specify the Global connection group by taking the default Global and click the OK button to finish the creation process, as shown in Figure 5-10.

Figure 5-10. *Create notebook*

The Connection field specifies how the notebook is connected to the back-end database where the query result is fetched. The connection group consists of a single connection type: Global. The Global connection group is created automatically when a new database is provisioned. It points to the ADW database that the creator of the notebook (JohnDev1 in this case) belongs to.

You can verify this configuration by logging in to OML as the admin account. On the pull-down menu of the left side of the login home page, click Connection Groups to go to the Connection Group: Global page (see Figure 5-11).

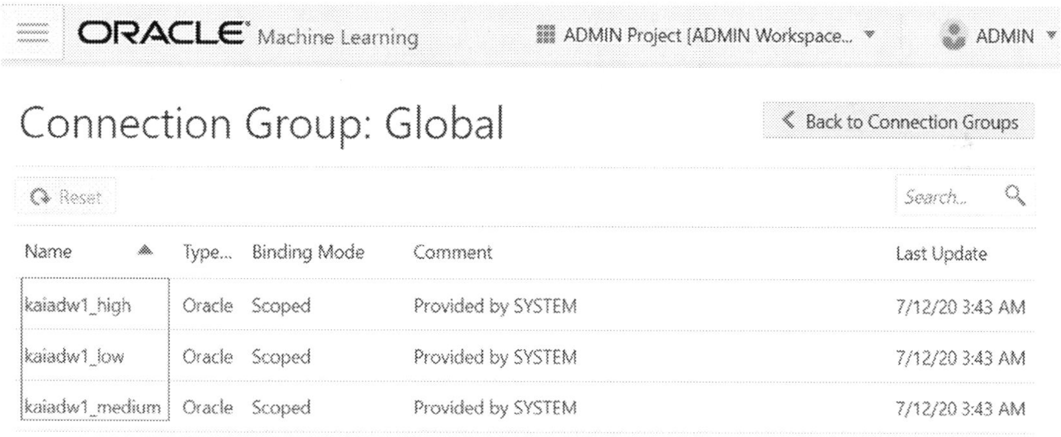

Figure 5-11. *Default global connection group*

This Global connection group consists of three database services (high, medium, and low) that point to the ADW database.

Specifying Interpreter Bindings and Connection Groups

You can write SQL statements or PL/SQL scripts in a notebook. You need to understand how these SQL statements and PL/SQL scripts are interpreted and executed. The answer to this question is related to the interpreter binding of the notebook. A Zeppelin notebook is composed of paragraphs that can contain text, SQL statements, and scripts (PL/SQL). These different contents in paragraphs need different interpreters to process. Usually, the name of the interpreter is set at the top of the paragraph as follows.

- %md: Markdown language using plain-text formatting syntax that can be converted to HTML format

- %sql: Runs a single SQL statement

- %script: Runs a set of SQL statements or PL/SQL scripts

Oracle Database provides the SQL interpreter and script interpreter on the back end to process the data processing language such as SQL and PL/SQL. For a notebook to use these interpreters, you need to bind it to an interpreter to fetch data from the database or any data source. Oracle Machine Learning Notebooks must set the interpreter binding to connect the SQL and PL/SQL interpreters within an ADW Database interpreter group.

A notebook has an interpreter setting specification that contains an internal list of bindings that determine the order of the interpreter bindings in an interpreter group. The names of the interpreters are in a databasename_servicename format; for example, kaiadw1_low %sql (default), %script in Figure 5-12. You can check the order of interpreter binding by clicking the gear icon, ⚙, in the upper-right corner of the screen.

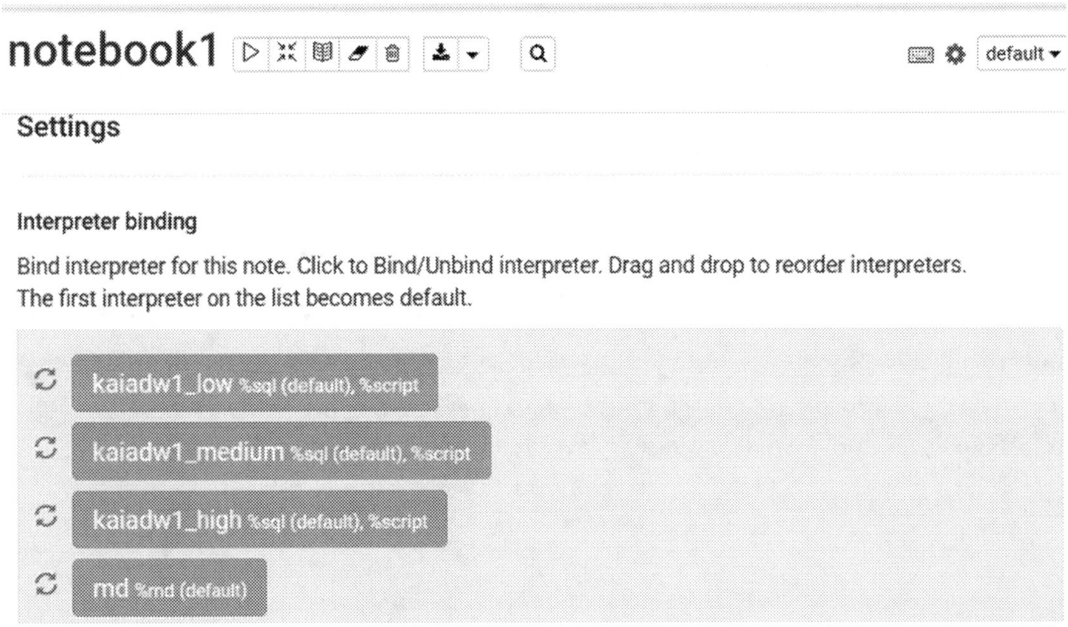

Figure 5-12. Interpreter binding

Figure 5-12 shows an example of the default order of interpreter bindings in Oracle Database interpreter groups.

- `kaiadw1_low` `%sql` (default), `%script`: It provides the least level of resources for each SQL statement. But it can support the maximum of the concurrent SQL statements. This interpreter has low priority and is listed as a default option. This interpreter can be used for both the SQL statement and PL/SQL script.

- `kaiadw1_medium` `%sql` (default), `% script`: It provides a lower level of resources of each SQL statement. But it can support more concurrent SQL statements. This interpreter can be used for both the SQL statement and PL/SQL script.

- `kaiadw1_high` `%sql` (default), `%script`: It provides the highest level of resources of each SQL statement. This interpreter is for the highest performance by supporting a minimum number of concurrent SQL statements. This interpreter can be used for both the SQL statement and PL/SQL script.

- `Md` `%md` (default): It provides plain text formatting syntax to be converted to HTML format.

You can change the order by dragging and dropping interpreter bindings. The first interpreter binding on the list is the default selection. You can click the bind/unbind interpreter for this notebook. You also can change the interpreter binding for any specific paragraph.

A notebook's connection group defines a group of connections to these interpreters. Since SQL and PL/SQL interpreters are used within an Oracle Database interpreter group, the connection group is also a collection of database connections. The Global Connection group is the default, as shown in Figure 5-13. The connection group is Global.

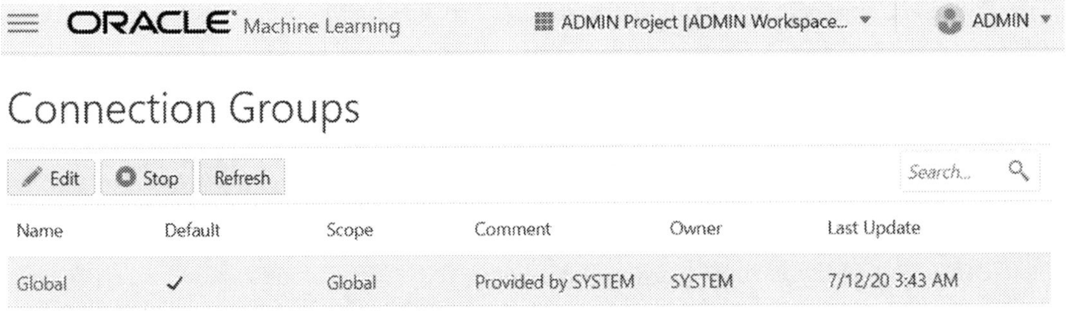

Figure 5-13. *Connection groups*

This Global Connection group is created automatically when a new ADW database is created. This Global Connection Group has the connection type as Global and a list of database connections as compute resources. Clicking the word Global shows the database connections listed as Compute resources, such as kaiadw1_high, kaiadw1_medium, and kaiadw1_low, as shown in Figure 5-14. You probably have noticed that these database connection names are also shown in the interpreter bindings in Figure 5-12. These connections indicate the connections to the interpreter bindings with the notebook.

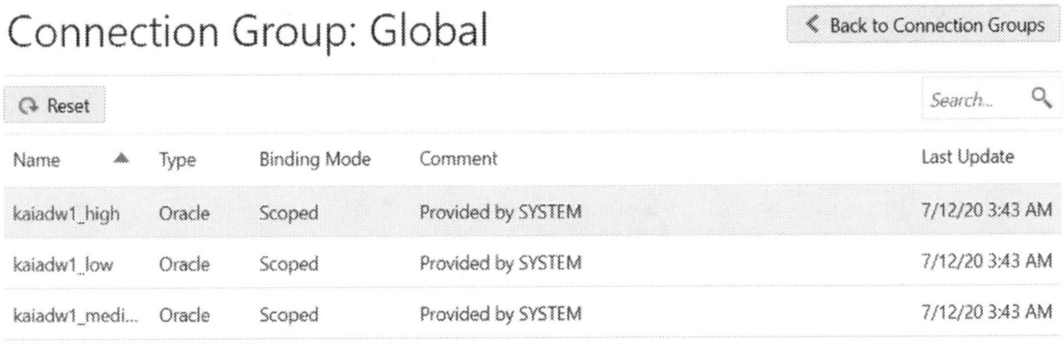

Figure 5-14. *The database connections of the Global Connection Group*

When a notebook uses the kaiadw1_low %sql interpreter binding, it uses the kaiadw1_low %sql interpreter to interpret the SQL statements. It also connects to the database using the kaiadw1_low database service.

Once a notebook is created with the interpreter binding settings, you can write paragraphs that contain SQL statements or PL/SQL scripts for data analysis, data discovery, and data visualization. You can run one or all the paragraphs of code by clicking the run icon, ▷. The result can be exported by clicking the download icon, ⬇.

Figure 5-15 shows that you can write and execute a SQL query statement in the notebook.

Figure 5-15. *Edit a notebook*

Running SQL Scripts and Statements

Usually database application development environment allows to write or execute both individual SQL statements and SQL or PL/SQL scripts. Let's discuss how Oracle Machine Learning Notebook environment supports these two.

Create and Execute SQL Scripts in a Notebook

With SQL Developer or the SQL*Plus tool, you can write and execute SQL scripts in Oracle Machine Learning Notebooks. You can put several SQL statements or PL/SQL blocks in a script, run the script in a notebook, and save the script as a JSON file.

To create a SQL script, on the OML home page, click Run SQL Scripts to open SQL Scratchpad. Then type **%script** and press Enter to open the next line to enter and edit a SQL script. Once you finish editing the script, you can click the ▷ button to run the script in the back-end database.

Figure 5-16 shows creating a sample SQL script that calls the CREATE_MODEL procedure. This is the ML4SQL procedure for machine learning model creation.

SQL Script Scratchpad ⏵ ✖ 📖 ✏ 🗑 ⬇ ▾

```
%script
BEGIN DBMS_DATA_MINING.DROP_MODE('CUSTOMER_CREDIT_MODEL');
EXCEPTION WHN OTHERS THEN NULLL;END;
/
BEGIN
    DBMS_DATA_MINING.CREATE_MODEL(
    MODEL_NAME          => 'CUSTOMER_CREDIT_NODEL'
    MINING_FUNCTION     => DBMS_DATA_MINING.ASSOCIATION,
    DATA_TABLE_NAME     => 'CUSTOMER_CREDIT',
    CASE_ID_COLUMN_NAME => 'CUST_ID',
    SETTINGS_TABLE_NAME => 'CUSTOMER_CREDIT_SAMPLE_SETTINGS'
    );
END;
```

Figure 5-16. *Create SQL and PL/SQL script*

You also can export a SQL PL/SQL script to a .json file in your local system by clicking the ⬇ icon.

Run SQL Statements in a Notebook

With the Connection Group setting and interpreter binding with the back-end ADW database, Oracle Machine Learning Notebooks can be considered a web front-end development and GUI interface of the ADW database. You can write SQL statements in a notebook and have them interpreted and executed by the back-end ADW database, similar to Oracle client tools like SQL* Plus or SQL Developer, to connect to the ADW database on the back end.

After you log in to OML, click Run SQL Statements in the Quick Actions section, the SQL Query Scratchpad opens. On the SQL Query Scratchpad, type %sql and press Enter. You can enter a SQL statement in the next line. To run the SQL statement, click ▷ or press the Shift+Enter keys. The execution status of the SQL statement shows as FINISHED if it runs successfully or as an ERROR with any error messages (see Figure 5-17).

```
%sql                                                        FINISHED ▷ ⅹ 🔖 ⚙

create table ship_address (customer_id integer, address varchar2(300));
```
Took 0 sec. Last updated by JOHNDEV1 at August 05 2020, 11:29:28 PM. (outdated)

```
%sql                                                          ERROR ▷ ⅹ 🔖 ⚙
create table ship_address (customer_id integer, address varchar2(300));

ORA-00955: name is already used by an existing object
```
Took 0 sec. Last updated by JOHNDEV1 at August 05 2020, 11:31:04 PM.

Figure 5-17. *Execute SQL statement with return status*

Oracle Machine Learning Notebooks enable you to visualize the data. If you run a select statement to fetch data from the back-end database. For example, in select * from sh.customers, the statement is sent to the interpreter to be executed. It fetches the data from the back-end database and displays it in the output section below the select statement. The output session displays the data in the tabular format as default, as shown in Figure 5-18.

CUST_ID ⌄	CUST_FIRST_NA...⌄	CUST_LAST_NAM.⌄	CUST_GENDER ⌄	CUST_YEAR_OF ≡
35228	Abner	Kenney	M	1952
37006	Abner	Robbinette	M	1953
40561	Abner	Robbinette	M	1959
44116	Abner	Robbinette	M	1956
34359	Abner	Robbinette	M	1971
47895	Abner	Robbinette	M	1963
1451	Abner	Robbinette	M	1962
5006	Abner	Robbinette	M	1989

Figure 5-18. *Run a select statement in a notebook*

Work with Notebooks to Analyze and Visualize Data

After the result data of a SQL select statement is presented in the output section, you can create various graphical presentations based on the result data. When you click any of the icons shown in Figure 5-19, you get the corresponding graphical display of the result data. The icons, as shown from the left, represent table, bar chart, pie chart, area chart, line chart, scatter chart, and download to CSV/TSV, respectively.

Figure 5-19. *Graphical display icons*

These graphical representations of the data are critical features in data analysis and data visualization and are widely used in machine learning.

The settings icon in Figure 5-20 allows you to specify the presentation's X axis and Y axis. Click it to show all the field names in the select statement. You can drag and drop the fields into the three boxes: keys, groups, and value. If you select the bar chart as the graphical display, settings allow you to specify the following.

- The keys tab and the groups tab set fields for the graph's X axis.

- The values tab sets the graph's Y axis, which are a field's SUM, COUNT, AVG, MIN, or MAX.

Figure 5-20 shows an example of a bar chart.

- keys values of cust_credit_ limit field for X-axis

- Count of CUST_ID as Y-axis

With this setting, The BAR chart shows the distributions of the number of customers by each customer credit limit. For example, the first bar shows that there are 195 customers with a credit limit=1500.

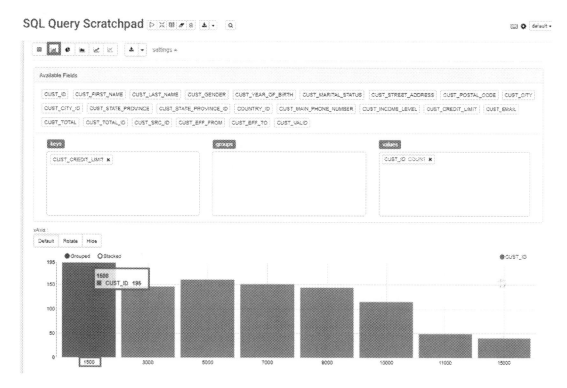

Figure 5-20. *Bar chart: number of customers by customer credit limit*

If the CUST_CREDIT_LIMIT field is moved to the groups tab, the unit on the bar chart's X axis changes, as shown in Figure 5-21.

Figure 5-21. *Bar chart: customer count by customer credit level*

You also can select the pie chart icon, 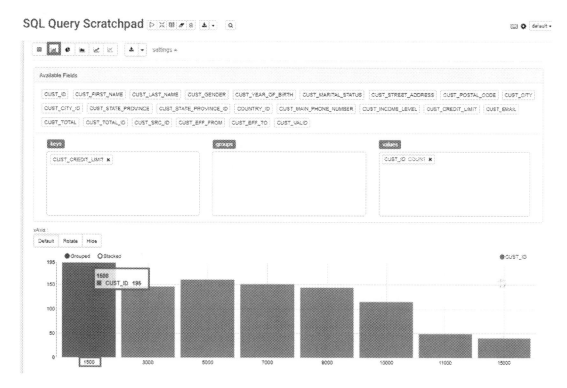. Figure 5-22 shows a pie chart of credit limits by the number of customers.

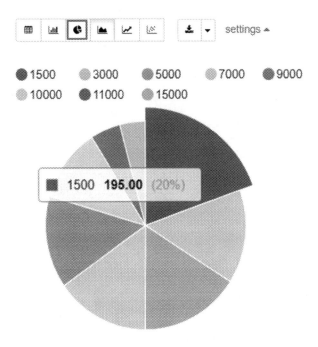

Figure 5-22. *Pie chart*

If you select the area chart icon or the line chart icon (see Figure 5-19) with the CUST_YEAR_OF_BIRTH field as the keys and CUST_ID count as the values, you get an area chart or a line chart that shows customers' birth year distribution, as shown in Figure 5-23.

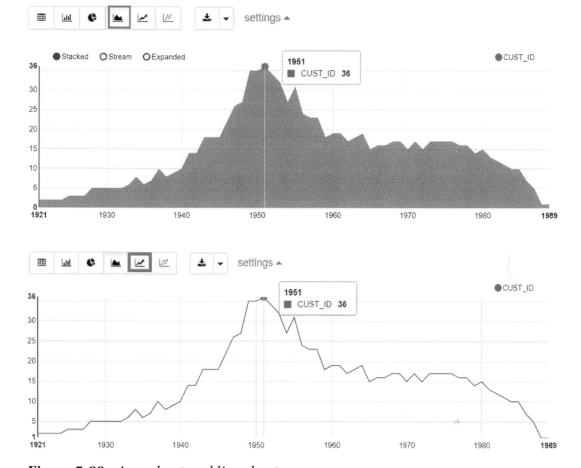

Figure 5-23. *Area chart and line chart*

The last icon is the scatter chart, which uses dots to represent values for two variables in the X axis and the Y axis. For example, the scatter chart in Figure 5-24 shows the relationship and data distribution of CUST_YEAR_OF_BIRTH (X axis) and CUST_INCOME_LEVEL (Y axis).

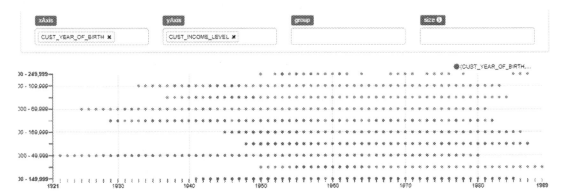

Figure 5-24. *Scatter chart*

Summary

This chapter has introduced Oracle Machine Learning as part of the built-in toolset of the Oracle ADW application environment. To access OML, the ADW database admin needs to create a new user or grant an existing user in ADW with the OML developer (OML developer) role. With this role, this user can log in to OML. After the login, this user is provided the default workspace and a default project. This user can also create a new workspace and project. The user can create a notebook after selecting the workspace and project—either the defaults or newly created ones. A notebook uses its connection setting to connect to the ADW database and uses the interpreter binding settings to connect to the SQL and PL/SQL interpreter of the ADW database to process the SQL statements and SQL scripts in the notebook.

You can write SQL scripts in a notebook and execute them in the back-end ADW database. This allows us to run OML4SQL in Oracle Machine Learning Notebooks. You also can write SQL statements in a notebook and have them executed in the back-end ADW database. A SQL query statement can fetch the result data from the ADW database and display them in the notebook. You can display the results in various graphical formats by selecting different display icons and X-axis and Y-axis settings. These formats include table, bar chart, pie chart, area chart, line chart, and scatter chart. These graphical representations of the result data are the keys to analyzing and visualizing data in OML projects.

The next chapter discusses how to use the OML environment.

CHAPTER 6

Building Machine Learning Models with OML Notebooks

Chapter 4 discussed the Oracle Autonomous Database environment and explored how to perform some essential tasks in Oracle Autonomous Database such as database provisioning, data loading, and database schema/table importing. Chapter 5 focused on a key component of Oracle Autonomous Database, Oracle Machine Learning, and discussed how to run SQL statements and SQL and PL/SQL scripts with Oracle Machine Learning Notebooks and how to visualize data fetched from Oracle Autonomous Database. Through these key features, Oracle Machine Learning (OML) in Oracle Autonomous Database provides an excellent platform for various machine learning development projects. This chapter is dedicated to exploring how to leverage this platform for machine learning projects.

In addition to Oracle Machine Learning Notebooks, Oracle Autonomous Databases such as Autonomous Data Warehouse (ADW) and Autonomous Transaction Processing (ATP) provide full support of Oracle Machine Learning for SQL, also called OM4SQL, through its Oracle database engine. OML4SQL as a component of Oracle Database is implemented in the Oracle Database kernel. It is included in the most common Oracle database license options such as SE2, EE, EE-ES, DBCS SE, DBCS EE, DBCS EE-HP, DBCS EE-EP, and ExaCS.

The machine learning capability of OML4SQL makes it possible for us to do in-database machine learning in Oracle databases running in your own infrastructure and Oracle Autonomous Database. Through Oracle Machine Learning Notebooks, you can use all the OML4SQL functionalities in Oracle Autonomous Database with OML4SQL programming interfaces in the PL/SQL API and SQL language operators and functions.

© Heli Helskyaho, Jean Yu, Kai Yu 2021
H. Helskyaho et al., *Machine Learning for Oracle Database Professionals*,
https://doi.org/10.1007/978-1-4842-7032-5_6

This chapter covers the following.

- Oracle Machine Learning overview

- Machine learning process flow

- Oracle Machine Learning for SQL

- An example machine learning project

Oracle Machine Learning Overview

Machine Learning tasks usually involve building intelligent systems that learn or improve performance based on a large set of data. Machine Learning can help solve various types of business problems. The following are some of the business problems that need machine learning.

- **Pattern discovery**: A machine learning system automatically discovers certain patterns or trends by searching a large set of data. Machine learning creates a model based on the existing data and then applies the model to new data.

- **Prediction**: A machine learning system can predict the probability of a future event based on some past data. In this case, machine learning creates a model based on the past data, then applies the model to future data to predict with associated probability (namely, the likelihood) that the prediction is true.

- **Grouping**: One form of machine learning is finding a natural grouping of data, such as identifying a segment of customers within a specified credit score range.

- **Actionable Information**: Machine learning can generate an actionable decision from a large set of data. For example, a machine learning system can create a model to assign the credit card limit to customer credit card applications.

Supervised Learning and Unsupervised Learning

Machine learning functions specify a class of problems that can be modeled and solved with machine learning. There are two general categories of these machine learning functions: supervised machine learning and unsupervised machine learning.

The supervised machine learning process is directed by a previously known target output and its dependent input attributes. The model establishes the target value as a function of a set of independent attributes based on a known data set. The building of this supervised learning model involves model training, model testing, and model scoring.

- **Model training**: This process analyzes a set of case data on which the target and attributes are known. In this training process, the model learns the target as a function of attributes. For example, a model training process establishes the customer credit score as a function of the attributes such as wealth, income, professions, family size, and job history based on a data set of known credit scores, and these attribute values.

- **Model testing**: This process applies the model to a known test data set and compares the target values generated from the model with the known target values to determine the model accuracy. In the last example, the model is applied to the test data and generates the target values: customer credit scores. The generated customer credit scores are compared with the real known customer credit scores in the test data.

- **Model scoring**: This process applies the model to the actual data (called *scoring data*) to predict the unknown target value for the scoring data. In the last example, the model scoring process applies the model to new customers. It predicts the credit scores of these new customers based on their wealth, income, professions, family size, job history.

Oracle Machine Learning supports the following supervised learning functions.

- **Attribute Importance**: This function identifies and ranks the attributes that are more important in predicting a target attribute; for example, ranking attributes such as income and occupation to determine a customer's credit score. Ranking by importance is useful for selecting those more important attributes in model training for classification and regression models. Oracle Machine Learning does not support the model scoring operation (applying model) for the Attribute Importance function.

- **Classification**: This function determines which category an object belongs to. For example, for a given demographic and personal data about a set of customers, predict whether a customer will buy certain products.

- **Regression**: This predicts a continuous valued attribute associated with an object. For example, based on customers' demographic data, predict how much a customer will spend during a sales event.

The unsupervised learning process is not directed. It is used when you do not know what the target output should be. In unsupervised learning, there are no previously known results to train the model and test the model. Unsupervised learning can be used for descriptive or predictive purposes. Oracle Machine Learning supports the following unsupervised learning functions.

- **Anomaly Detection**: It finds whether cases meet the characteristics of normal cases, such as determining whether a customer's purchasing pattern is different from the normal among similar customers.

- **Association Rules**: It identifies the pattern of association within the data, such as the products that customers tend to buy together.

- **Clustering**: It groups items that are similar to one another; for example, it groups the voters that will vote for a certain candidate.

- **Feature Extraction**: It creates new attributions using linear combinations of the original attributions.

Oracle Machine learning supports the model scoring operation for clustering and feature extraction but does not support the scoring operation for association rules. Association models cannot be applied to separated data. Instead, they can explain or return the rules about how the items and events are associated with each other.

To build machine learning models, Oracle Machine Learning supports at least one or more mathematical procedures called *algorithms* for each machine learning function. Table 6-1 lists the machine learning algorithms that Oracle Machine Learning supports for their corresponding machine learning functions.

Table 6-1. *Oracle Machine Learning Functions and Algorithms*

Function	Type	Algorithms
Attribute Importance	Supervised Learning	Minimum Description Length
Classification	Supervised Learning	Decision Tree, Explicit Semantic Analysis, Naive Bayes, Random Forest,
Classification and Regression	Supervised Learning	Generalized Linear Model Neural Network Support Vector Machine XGBoost
Anomaly Detection	Unsupervised Learning	CUR Matrix Decomposition, Multivariate State Estimation Technique – Sequential Probability Ratio Test, One-Class Support Vector Machine
Association Rules	Unsupervised Learning	Apriori, *k*-Means,
Clustering	Unsupervised Learning	Expectation Maximization
Feature Extraction	Unsupervised Learning	Explicit Semantic Analysis

Detailed descriptions of each algorithm are in *Oracle Machine Learning for SQL Concepts, 21c,* F31930-05, June 2021, Oracle Corporation. (URL: `https://docs.oracle.com/en/database/oracle/machine-learning/oml4sql/21/dmcon/index.html`).

Machine Learning Process Flow

A typical machine learning process consists of the following main phases.

- **Business problem definition**: This initial phase involves obtaining the business problem requirements, assessing the goals of the machine learning project. To understand the project objectives and requirements, you can define the machine learning problem and develop a project plan.

- **Data gathering and preparation**: Once the business requirement is established, the next step is to ensure that quality data is available for the business problem. This phrase includes data collection, data cleaning and quality check, data exploration, and data analysis, and data sampling and data transformation. It also includes feature or attribution selection and splitting the data set into training data and test data.

- **Model training**: This phrase uses the selected machine learning algorithm to create the machine learning model by understanding the pattern of data and derive the insights out of the pattern. This process trains the model based on the train data set. In supervised learning, the model establishes the target attribute as a function of its dependent attributes.

- **Model evaluation and testing**: A newly created model needs to be tested for its accuracy and evaluated to see if this model can meet the requirements of the business problem. One way to test the model is to apply the model to the test data and then compare the target values generated from the model with the real target values of the test data.

- **Model scoring and application**: After the model passes the model evaluation and testing, the model can be applied to new data to generate the new target values which are previously unknown. A model application can also be deployed into the business application environment. For example, a model scoring of predicting customer credit scores can be deployed in the customer credit card approval application.

The next sections discuss how to perform these major phases of the machine learning process in the Oracle Machine Learning environment with Oracle Autonomous Database.

Oracle Machine Learning for SQL

Oracle Autonomous Database such as ADW supports in-database machine learning. All the machine learning functionalities are implemented inside an Oracle database, which is the ADW in this case. This in-database machine learning is accomplished with Oracle Machine Learning for SQL (OML4SQL).

OML4SQL is implemented as part of the Oracle Database kernel and available to use in on-premise Oracle Databases and Oracle Autonomous Databases such as ADW. Therefore, when you provision an ADW in Oracle Cloud Infrastructure, the ADW is fully equipped with OML4SQL in the database kernel. OML4SQL models are stored as first-class database objects that can be exported. When you use OML4SQL to perform machine learning tasks, you can take advantage of maximal scalability and Oracle Database system resources.

As shown in Table 6-1, OML4SQL provides a set of in-database machine algorithms that can implement various machine learning functions, such as classification, regression, feature selection, clustering, anomaly detection, feature extraction. These machine learning algorithms can directly work on the data such as star schema data, transactional data, and other forms of unstructured data in the ADW. Therefore, you do not need to copy and convert a large set of data from the ADW to another machine learning platform. By eliminating the data movement and only keeping one copy of the data in the ADW, the data set is always protected by the extensive security features of the ADW. Only the users with specific database privileges can access the data and perform related OML4SQL tasks.

OML4SQL PL/SQL API and SQL Functions

You can access OML4SQL functionalities in Oracle Database through its application programming interfaces: the PL/SQL API and SQL language operators/functions.

The PL/SQL API consists of a set of PL/SQL packages for building and maintaining models. One frequently used PL/SQL package is the DBMS_DATA_MINING package, which contains the routines for building, testing, and maintaining machine learning models. SQL language operators and SQL functions are widely used in data preparation and model scoring phases. Let's discuss different phases of in-database machine learning with Oracle database and the usage of these OML4SQL API and SQL functions in these phases.

Data Preparation and Data Transformation

Usually, a machine learning project accesses the data defined in a single table or view in Oracle Database. The first step of data preparation is to make sure that the data is stored in a database table or a view, and the information for each case is stored in a single row of the table or view. If the information resides in multiple tables, you can create a view that joins these tables.

Split Data

You can use database views to split the data sets. For classification and regression, two data sets should be prepared. One data set (build data) for training the model and another data set (test data) for testing models. Both data sets should be derived from the same data. For example, you can create two views that select random samples of 60% and 40% of the rows for the build data set and the test data set, as shown in Listing 6-1.

Listing 6-1. Split Data with Views

```
create or replace view credit_scoring_build_v as select *
from admin.credit_scoring sample (60);

create or replace view credit_scoring_test_v as select *
from admin.credit_scoring sample (40);
```

Data Transformation

Data analysts often need to analyze data during data preparation by performing some data transformation before the data can build a model. These data transformations include binning numeric data, handling missing values, and selecting the top N attributes (feature selection) used for model building. A data transformation is done through a SQL expression that modifies one or more columns of the data set (table). OML4SQL automates most of these transformations with Automatic Data Preparation (ADP).

Through ADP, these transformations are embedded in the model and automatically executed when the model is applied. You also can specify additional transformations with SQL expression as input to the model creation process.

Transformation Expressions

OML4SQL API provides the DBMS_DATA_MING_TRANSFORM package for transformation operations that are commonly used in machine learning. This package offers two approaches to implement transformations.

One method is to create a list of transformation expressions and pass the list to the CREATE_MODEL procedure. This transform list can be created with the SET_TRANSFORM procedure. The following is an example of this approach. A list of transformation expressions v_xlst is declared and built with the dbms_data_mining_transform.set_transform procedure. Then this list is passed to DBMS_DATA_MINING. CREATE_MODEL model creation procedure through the xform_ list parameter as shown in Listing 6-2.

Listing 6-2. Pass a List of Transformation Expressions Through the Model Creation Procedure

```
declare
  v_xlst dbms_data_mining_transform.TRANSFORM_LIST;

BEGIN
  dbms_data_mining_transform.set_transform(v_xlst,
    'HOME_THEATER_PACKAGE', NULL, 'HOME_THEATER_PACKAGE',
    'HOME_THEATER_PACKAGE', 'FORCE_IN');

  DBMS_DATA_MINING.CREATE_MODEL(
    model_name          => 'GLMR_SH_Regr_sample',
    mining_function     => dbms_data_mining.regression,
    data_table_name     => 'mdb_rdemo_v',
    case_id_column_name => 'cust_id',
    target_column_name  => 'age',
    settings_table_name => 'glmr_sh_sample_settings',
    xform_list          => v_xlst);
END;
```

Binning Transformations

The second approach is to create a view that implements the transformation and pass the name of the view to the CREATE_MODEL Procedure. The following example shows how to create a view called 'mining_data_bin_view' that implements the binning transformation, as shown in Listing 6-3.

Listing 6-3. Binning Transformation

```
dbms_data_mining_transform.create_bin_num(
        bin_table_name     => 'bin_num_tbl');

 dbms_data_mining_transform.insert_autobin_num_eqwidth(
        bin_table_name     => 'bin_num_tbl',
        data_table_name  => 'CREDIT_SCORING_100K_V',
        bin_num            => 5,
        max_bin_num        => 10,
        exclude_list       => );

dbms_data_mining_transform.COLUMN_LIST('CUSTOMER_ID'));

dbms_data_mining_transform.xform_bin_num(
        bin_table_name     => 'bin_num_tbl',
        data_table_name  => 'CREDIT_SCORING_100K_V',
        xform_view_name  => 'mining_data_bin_view');
```

This binning transformation, also called *discretization*, is one of the commonly needed transformations in machine learning. It is a technique for reducing the cardinality of data. By binning groups of related values in bins, it reduces the number of distinct values of an attribute of the data set. ADP automatically bins numeric attributes using either default or user-customizable binning strategies, such as equal width, equal count, and user-defined bins. Once the binning view is created, this SQL query runs on the bin view mining_data_bin_view in OML Notebooks (see Listing 6-4).

Listing 6-4. mining_data_bin_view Query

```
Select customer_id, age, income, education_level from ming_data_bin_view;
```

By selecting AGE in the Key field and CUSTOMER_ID COUNT in the values files of the notebook, Figure 6-1 shows a histogram of the numbers of customers by each AGE bin group.

Figure 6-1. *Histogram of the numbers of customers by AGE bin group*

Model Creation

OML4SQL provides the DBMS_DATA_MINING.CREATE_MODEL procedure for machine learning model creation. As shown in the following code, the CREATE_MODEL procedure takes a set of parameters as inputs to create a machine learning model, as shown in Listing 6-5.

Listing 6-5. Model Creation Procedure

```
PROCEDURE CREATE_MODEL(
      model_name IN VARCHAR2,
      mining_function IN VARCHAR2,
      data_table_name IN VARCHAR2,
      case_id_column_name IN VARCHAR2,
      target_column_name IN VARCHAR2 DEFAULT NULL,
      settings_table_name IN VARCHAR2 DEFAULT NULL,
      data_schema_name IN VARCHAR2 DEFAULT NULL,
      settings_schema_name IN VARCHAR2 DEFAULT NULL,
      xform_list IN TRANSFORM_LIST DEFAULT NULL);
```

Therefore, before calling the CREATE_MODEL procedure, you need to prepare these parameters.

1. Decide the name of the machine learning function and algorithm for the mining_function parameter. The various machine learning functions and algorithms that OML4SQL supports are shown in Table 6-1.

2. Prepare the model training data set for the data_table_name parameter. For example, we created a view that extracts 60% of the data set for model training.

3. Identify the target attribute. For example, supervised learning predicates the target attribute as dependent on other attributes. The target attribute is passed as CREATE_MODEL procedure through the target_column_name parameter.

4. Specify model settings. These are the configuration settings for machine learning model creation. These model settings can be specified in a settings table with two columns: setting_name and setting_values. Some of the settings are for machine learning model functions, and some of the settings are for machine learning algorithms. The settings table is passed to the CREATE_MODEL procedure through the settings_table_name parameter.

5. Specify the transformation list. This list of transformation expressions specifies the required attribute value transformation for the model creation. This list is passed to the model creation procedure through xform_list.

The CREATE_MODEL procedure uses the data to create a machine learning model with the model function/algorithm, and other parameters passed to the procedure. The created model is a database table with a name specified in the model_name parameter in the procedure.

You can use this sample code to call the model creation procedure, as shown in Listing 6-6.

Listing 6-6. Model Creation Procedure Example

```
declare
  v_xlst dbms_data_mining_transform.TRANSFORM_LIST;

BEGIN
  dbms_data_mining_transform.set_transform(v_xlst,
    'HOME_THEATER_PACKAGE', NULL, 'HOME_THEATER_PACKAGE',
    'HOME_THEATER_PACKAGE', 'FORCE_IN');
  DBMS_DATA_MINING.CREATE_MODEL(
    model_name           => 'GLMR_SH_Regr_sample',
    mining_function      => dbms_data_mining.regression,
    data_table_name      => 'mdb_rdemo_v',
    case_id_column_name  => 'cust_id',
    target_column_name   => 'age',
    settings_table_name  => 'glmr_sh_sample_settings',
    xform_list           => v_xlst);
END;
```

Model Evaluation

Once a model is created with the model training process based on the training data set, you can evaluate the accuracy of the model. For supervised machine learning models such as the classification model, since the known target values are in the data set, you can apply the newly created model to the test data and then compare the model's predictions with the target values in the test data.

This model evaluation consists of two steps: model application and result comparison.

Model Application

This step applies the model to the test data to generate the prediction results. The APPLY procedure is used in the DBMS_DATA_MINING package to apply the model to the test data table and write the model results to a result table. The syntax of the DBMS_DATA_MINING.APPLY procedure is shown in Listing 6-7.

Listing 6-7. Model Application Procedure

```
DBMS_DATA_MINING.APPLY (
        model_name IN VARCHAR2,
        data_table_name IN VARCHAR2,
        case_id_column_name IN VARCHAR2,
        result_table_name IN VARCHAR2,
        data_schema_name IN VARCHAR2 DEFAULT NULL);
```

The following are the procedure's parameters.

- `model_name`: The name of the model

- `data_table_name`: The name of table or view containing the data to be tested

- `case_id_column_name`: The name of the case identifier column

- `result_table_name`: The name of the table in which to store the applied results

- `data_schema_name`: The name of the schema containing the data to be scored

By executing the following procedure in OML Notebooks, the 'N1_CLASS_MODEL' model is applied to the 'N1_TEST_DATA' test data table, which produces the results that are stored in the 'N1_APPLY_RESULT' table (see Listing 6-8).

Listing 6-8. Model Application Procedure Example

```
DBMS_DATA_MINING.APPLY(
        model_name            => 'N1_CLASS_MODEL',
        data_table_name       => 'N1_TEST_DATA',
        case_id_column_name => 'CUSTOMER_ID',
        result_table_name     => 'N1_APPLY_RESULT');
```

Result Comparison

This step takes the results from the model application to compare with the real target values. As one way to make this comparison, the COMPUTE_LIFT procedure in the DBMS_DATA_MINING package is designed to compute lift, which is a test metric for binary classification models. The COMPUTE_LIFT compares the predictions generated by the models with the actual target values in the test data set. Lift measures how much better the model's prediction is than random chance. The syntax of this procedure is shown in Listing 6-9.

Listing 6-9. COMPUTE_LIFT Procedure

```
DBMS_DATA_MINING.COMPUTE_LIFT (
    apply_result_table_name =>      'N1_APPLY_RESULT'
    target_table_name       => 'N1_TEST_DATA'
    case_id_column_name      => 'CUSTOMER_ID'
    target_column_name       => 'CREDIT_SCORE_BIN'
    lift_table_name          =>  'N1_LEFT_TABL
    positive_target_value    =>  'GOOD CREDIT'
    score_column_name        =>      'PREDICTION',
    score_criterion_column_name =>'PROBABILITY',
    num_quantiles               => 100)
```

The parameters in this procedure include the following.

- apply_result_table_name: The name of the table in which to store the apply results

- target_table_name: The name of the table that stores the test data sets

- case_id_column_name: The name of the case identifier column.

- target_column_name: The name of the target column

- lift_table_name: The name of the lift table

- positive_target_value: The value of the positive target

- score_column_name: The name of the column representing the score in the apply results table; default value 'PREDICTION'

- score_criterion_column_name: The name of the column representing the ranking factor for the score in the apply results table; default value 'PROBABILITY'

- num_quantiles: The number of quantiles required in the lift table

In Listing 6-10, N1_LIFT_TABLE is queried to review the model evaluation result.

Listing 6-10. Query the Lift Table

```
Select QUANTITLE_NUMBER, GAIN_CUMMULATIVE from N1_LIFT_TABLE;
```

In the OML Notebooks graphical display options, these query results are displayed as a gain and lift chart. This chart measures the effectiveness of a classification model by calculating the ratio between the results obtained with and without the model. This chart provides a visual aid for evaluating the performance of classification models. As shown in Figure 6-2, the chart consists of a lift curve of the model and a baseline (random). The greater the area between the lift curve and the baseline, the better the model is.

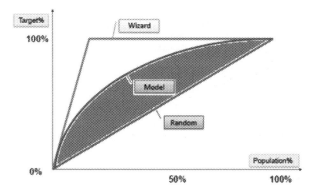

Figure 6-2. *Gain and lift chart*

In this example, the query results on NI_LIFT_TABLE are displayed as a gain and lift chart, as shown in Figure 6-3. The size of the area between the curve and the straight baseline indicates the effectiveness of the model.

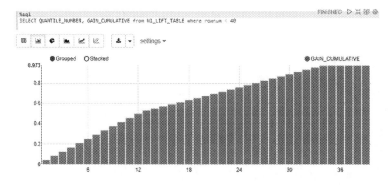

Figure 6-3. *The gain and lift chart shows the model's effectiveness*

Model Scoring and Model Deployment

Model scoring, also called *model application,* is a process to apply models to new data. Once a model is created and evaluated based on the existing known data set, the model can be applied to new data to predicate new target results.

With OML4SQL, model scoring is performed in a database by SQL language functions. The PREDICTION function is used for classification, regression, and anomaly detection. The following query is an example to show how to use the PREDICTION function to apply a model, as shown in Listing 6-11.

Listing 6-11. Use Prediction Function

```
SELECT cust_gender, COUNT(*) AS cnt, ROUND(AVG(age)) AS avg_age
FROM new_data_v
WHERE PREDICTION(dt_sh_clas_sample USING *) = 1
GROUP BY cust_gender
ORDER BY cust_gender;
```

The PREDICTION uses a USING clause to specify which attributes to use for scoring. In this example, USING * indicates that all the attributes are selected as predictors. You also can just specify some of the attributes as predictors. In the query shown in Listing 6-12, customers' gender, marital status, occupation, and income level are used as predictors.

Listing 6-12. An Example of the Prediction Function

```
SELECT cust_gender, COUNT(*) AS cnt, ROUND(AVG(age)) AS avg_age
FROM new_data_v
WHERE PREDICTION(dt_sh_clas_sample USING *) = 1
GROUP BY cust_gender
ORDER BY cust_gender;

SELECT cust_gender, COUNT(*) AS cnt, ROUND(AVG(age)) AS avg_age
FROM mining_data_apply_v
WHERE PREDICTION(dt_sh_clas_sample USING
cust_gender,cust_marital_status,
occupation, cust_income_level) = 1
GROUP BY cust_gender
ORDER BY cust_gender;
```

The query also supports missing attributes and expressions as predictors. This query uses customer gender, marital status, and occupation as predictors. It uses an expression to assume that all the customers are in the income level of $300,000 or higher, as shown in Listing 6-13.

Listing 6-13. An Example of Prediction Function with Expression

```
SELECT cust_gender, COUNT(*) AS cnt, ROUND(AVG(age)) AS avg_age
FROM mining_data_apply_v
WHERE PREDICTION(dt_sh_clas_sample USING
cust_gender, cust_marital_status, occupation,
'L: 300,000 and above' AS cust_income_level) = 1
GROUP BY cust_gender
ORDER BY cust_gender;

C CNT AVG_AGE
- ---------- ----------
F 30 38
M 186 43
```

PREDICTION_PROBABILITY is another SQL function that can apply the model. This function predicts the probability by applying the model to the new data. The query shown in Figure 6-4 applies the model to a single record to predicate the probability of having "good credit" for a new customer whose wealth level is Rich, income is $2000, and customer value segment level is Silver.

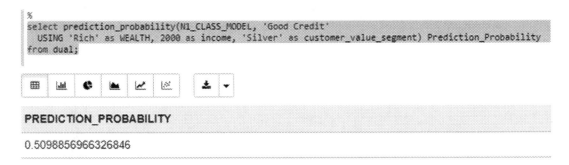

Figure 6-4. *Probability prediction*

Model scoring also can be performed by the DBMS_DATA_MINING.APPLY procedure. In the model testing session, we discussed how to use this APPLY procedure to test data to get the predicted results. The method also works for applying the model to new data. The APPLY procedure generates the same results as scoring with the SQL scoring functions and creates a result output table with CASE_ID, PREDICATION, PROBABILITY columns for classification, anomaly detection, clustering machine learning functions. For regression function, this result table has CASE_ID, PREDICATION columns; For the feature extraction function, this table has CASE_ID, FEATURE_ID, and MATCH_QUALIFY columns.

In Listing 6-14, the APPLY procedure applies the N1_CLASS_Model model to the new N1_NEW_DATA data set table with CUSTOMER_ID as the case_id. As a result, the procedure generates the N1_NEW_RESULT table.

Listing 6-14. Apply Model to New Data

```
Exec DBMS_DATA_MINING.APPLY ('N1_CLASS_MODEL', 'N1_NEW_DATA', 'CUSTOMER_
ID','N1_NEW_RESULT')
```

Listing 6-15 queries the results table to see the probability for each customer ID.

Listing 6-15. Query Probability

```
Select CUSTOMER_ID,PREDICATION,PROBABILITY      from N1_NEW_RESULT;
```

To leverage the results of the models, the models often are deployed in a production application environment. With the SQL functions and the DATA_MINING.APPLY for model scoring, OML4SQL supports model deployment in various application production environments with real-time and batch model scoring. The model applications may involve extracting model detailed results to produce reports or extending business intelligence by incorporating prediction results or exporting/importing the model from model training database to model scoring database environment.

An Example of Machine Learning Project

The remainder of this chapter looks at an example of building and applying a machine learning model in the ADW Oracle Machine Learning environment. We use the sample schema that is preloaded in the ADW database. All the SQL queries and PL/SQL scripts used for this example are documented in the code repository, which can be downloaded and executed in your own ADW OML environment.

Once you create an ADW database, it comes with the SH schema with the preloaded tables shown in Figure 6-5 with SQL Developer.

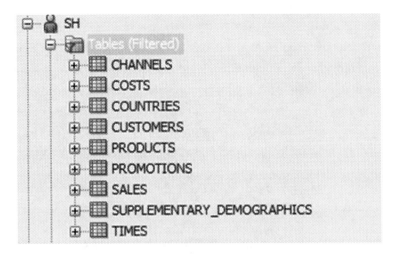

Figure 6-5. *SH schema in the ADW database*

Oracle Machine Learning Notebooks with the ADW have access to these tables. You can execute the SQL statements and PL/SQL scripts against the ADW database through OML Notebooks. The visualization capacity of OML Notebooks provides various graphical presentations based on the result data that are fetched from the ADW database through SQL queries.

Classification Prediction Example

This example uses the classification decision tree function to build a model that predicts whether a customer will likely buy a home theater package based on the customer's information. This supervised learning model is built based on the known sample data stored in the ADW database.

Data Preparation and Data Transformation

The example uses a single SUPPLEMENTARY_DEMOGRAPHICS table of the sample SH schema in the ADW database. The information of each case for model building is stored in a single row of this table. This table has 14 columns, as shown in Figure 6-6. Each column is corresponding to an attribute of cases.

	COLUMN_NAME	DATA_TYPE	NULLABLE	DATA_DEFAULT	COLUMN_ID	COMMENTS
1	CUST_ID	NUMBER	No	(null)	1	(null)
2	EDUCATION	VARCHAR2 (21 BYTE)	Yes	(null)	2	(null)
3	OCCUPATION	VARCHAR2 (21 BYTE)	Yes	(null)	3	(null)
4	HOUSEHOLD_SIZE	VARCHAR2 (21 BYTE)	Yes	(null)	4	(null)
5	YRS_RESIDENCE	NUMBER	Yes	(null)	5	(null)
6	AFFINITY_CARD	NUMBER(10,0)	Yes	(null)	6	(null)
7	BULK_PACK_DISKETTES	NUMBER(10,0)	Yes	(null)	7	(null)
8	FLAT_PANEL_MONITOR	NUMBER(10,0)	Yes	(null)	8	(null)
9	HOME_THEATER_PACKAGE	NUMBER(10,0)	Yes	(null)	9	(null)
10	BOOKKEEPING_APPLICATION	NUMBER(10,0)	Yes	(null)	10	(null)
11	PRINTER_SUPPLIES	NUMBER(10,0)	Yes	(null)	11	(null)
12	Y_BOX_GAMES	NUMBER(10,0)	Yes	(null)	12	(null)
13	OS_DOC_SET_KANJI	NUMBER(10,0)	Yes	(null)	13	(null)
14	COMMENTS	VARCHAR2 (4000 BYTE)	Yes	(null)	14	(null)

Figure 6-6. *SUPPLEMENTARY_DEMOGRAPHICS table structure*

The target attribute home_theater_package's value is 1 or 0 defines whether the customer will buy the home theater package. All the cases (rows in the table) have a known value of 0 or 1 for the target attribute (column).

Let's start exploring the data in the SUPPLEMENTARY_DEMOGRAPHICS table. To explore the target data distribution, execute the following query in OML Notebooks to fetch the data in Listing 6-16.

Listing 6-16. Fetch Data in OML Notebooks

```
select cust_id, HOME_THEATER_PACKAGE from sh.supplementary_demographics;
```

To visualize the data, adjust the display settings in the OML notebook as follows.

1. Click the bar chart icon and expand the settings.

2. Drag HOME_THEATER_PACKAGE to keys and CUST_ID to values. There should be nothing in groups.

3. Click CUST_ID in values and select count.

OML Notebooks shows the distributions of the target home_theater_package values between 0 and 1 by the CUST_ID count, as shown in Figure 6-7.

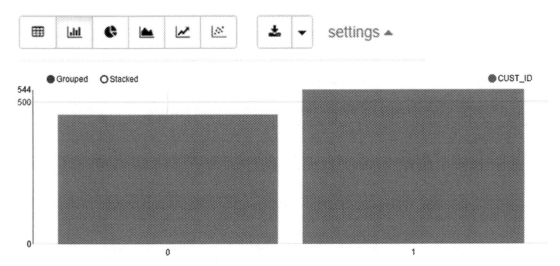

Figure 6-7. *Home_theater_package target value distribution*

Similarly, the query in Listing 6-17 shows the graphical distributions of the number of customers for YRS_RESIDENCE, Y_BOX_GAMES columns grouped by HOME_THEATER_PACKAGE column.

Listing 6-17. Fetch the Graphical Distribution Information

```
SELECT COUNT(CUST_ID) AS NUM_CUSTOMERS, Y_BOX_GAMES, YRS_RESIDENCE, HOME_
THEATER_PACKAGE
  FROM SH.SUPPLEMENTARY_DEMOGRAPHICS
  GROUP BY Y_BOX_GAMES, YRS_RESIDENCE, HOME_THEATER_PACKAGE
```

Figure 6-8 shows the process to display the distributions of the number of customers by group.

4. Click the bar chart icon and expand settings.

5. Drag Y_BOX_GAMES and YRS_RESIDENCE to keys. Drag HOME_
    THEATER_PACKAGE to groups.

6. Click NUM_CUSTOMERS in values and select SUM.

7. Click the Stacked button.

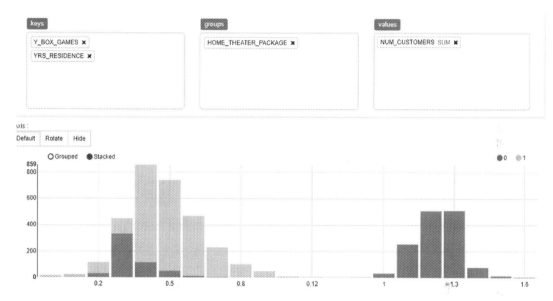

Figure 6-8. *Distributions of the number of customers by group*

To prepare the data, we created a view DEMOGRAPHICS_HTP_V based on
SH.SUPPLEMENTARY_DEMOGRAPHICS, as shown in Listing 6-18.

Listing 6-18. DEMOGRAPHICS_HTP_V View Creation

```
%sql
CREATE OR REPLACE VIEW DEMOGRAPHICS_HTP_V
AS (SELECT AFFINITY_CARD, BOOKKEEPING_APPLICATION,
    BULK_PACK_DISKETTES, CUST_ID,EDUCATION, FLAT_PANEL_MONITOR,
    HOME_THEATER_PACKAGE, HOUSEHOLD_SIZE,OCCUPATION,
    OS_DOC_SET_KANJI, PRINTER_SUPPLIES, YRS_RESIDENCE,
    Y_BOX_GAMES
    FROM SH.SUPPLEMENTARY_DEMOGRAPHICS);
```

Based on this view, the data is split randomly into two parts: 60% for training data (N1_TRAIN_DATA view) and 40% for testing data (N1_TEST_DATA), as shown in Listing 6-19.

Listing 6-19. Prepare Training and Testing Data with Views

```
CREATE OR REPLACE VIEW N1_TRAIN_DATA AS SELECT * FROM DEMOGRAPHICS_HTP_V
SAMPLE (60) SEED (1);

CREATE OR REPLACE VIEW N1_TEST_DATA AS SELECT * FROM DEMOGRAPHICS_HTP_V
MINUS SELECT * FROM N1_TRAIN_DATA;
```

Predicting Attribute Importance

OM4SQL provides the EXPLAIN function in the DBMS_PREDICTIVE_ANALYTICS package to analyze and create an attribute importance model. This function uses the Minimum Description Length algorithm to analyze the relative importance of attributes in predicting a target value. The EXPLAIN function in Listing 6-20 creates a new result table that saves a list of attributes ranked in the relative order of their impact on the prediction.

Listing 6-20. Explain Function

```
BEGIN
DBMS_PREDICTIVE_ANALYTICS.EXPLAIN(
   data_table_name          => 'DEMOGRAPHICS_HTP_V',
   explain_column_name      => 'HOME_THEATER_PACKAGE',
   result_table_name        => 'EXPLAN_RESULTS);
END;
```

As shown in Listing 6-20, the result table is EXPLAIN_RESULTS. Listing 6-21 is a simple SQL query that lists the top N (N=6) attributes .

Listing 6-21. Query's Top Six Attributes

```
select * from ai_explain_output_HOME_THEATER_PACKAGE where rownum<7;
```

Drag EXPLAINATORY_VALUE to value and select SUM. Drag ATTRIBUTE_NAME to keys. The bar graph shows the top six attributes ranked by the importance (EXPLAINATORY_VALUE) of predicting the target HOME_THEATER_PACKAGE value. A bar chart can display the top six attributes by importance, as shown in Figure 6-9.

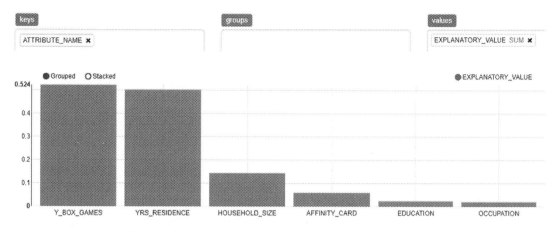

Figure 6-9. *Top six attributes by importance*

Model Creation

You can build a machine learning model with the training data in the N1_TRAIN_DATA
table by executing the PL/SQL script shown in Listing 6-22 in OML Notebooks.

Listing 6-22. Executing the Model Creation PL/SQL Script

```
%script
BEGIN
  DBMS_DATA_MINING.DROP_MODEL('HTP_CLASS_MODEL');
  EXCEPTION WHEN OTHERS THEN NULL;
END;
/
BEGIN
  EXECUTE IMMEDIATE 'CREATE TABLE N1_BUILDING_SETTINGS
     (setting_name varchar2(30), setting_value varchar2(4000))';
  EXECUTE IMMEDIATE 'INSERT INTO N1_BUILDING_SETTINGS  values
     (''PREP_AUTO'', ''ON'')';
  EXECUTE IMMEDIATE 'INSERT INTO N1_BUILDING_SETTINGS  values
     (''ALGO_NAME'', ''ALGO_DECISION_TREE'')';
  DBMS_DATA_MINING.CREATE_MODEL(
       model_name      => 'HTP_CLASS_MODEL',
       mining_function => 'CLASSIFICATION',
       data_table_name  => 'N1_TRAIN_DATA',
```

```
            settings_table_name => 'N1_BUILDING_SETTINGS',
            case_id_column_name => 'CUST_ID',
            target_column_name => 'HOME_THEATER_PACKAGE');
END;
```

This script creates a model stored in the HTP_CLASS_MODEL table.

As an alternative, the CREATE_MODEL2 procedure can be used to create a model. This procedure allows us to create a model without creating a persistent table or view of the input table, such as the N1_TRAIN_DATA. Instead, you can pass a query to the CREATE_MODEL2 procedure through the data_query parameter, as shown in Listing 6-23.

Listing 6-23. Passing a Query CREATE_MODEL2 Procedure

```
SELECT * FROM DEMOGRAPHICS_HTP_V SAMPLE (60) SEED (1)
```

The example code is in Listing 6-24.

Listing 6-24. An Example of Calling a CREATE_MODEL2 Procedure

```
%script
BEGIN DBMS_DATA_MINING.DROP_MODEL('HTP_CLASS_MODEL');
EXCEPTION WHEN OTHERS THEN NULL; END;
/
DECLARE
    v_setlst DBMS_DATA_MINING.SETTING_LIST;

BEGIN
    v_setlst('PREP_AUTO') := 'ON';
    v_setlst('ALGO_NAME') := 'ALGO_DECISION_TREE';

    DBMS_DATA_MINING.CREATE_MODEL2(
        model_name          => 'HTP_CLASS_MODEL',
        mining_function     => 'CLASSIFICATION',
        data_query          => 'SELECT * FROM DEMOGRAPHICS_HTP_V
                                SAMPLE (60) SEED (1)',
        set_list            => v_setlst,
        case_id_column_name => 'CUST_ID',
        target_column_name  => 'HOME_THEATER_PACKAGE');
END;
```

This example also shows how to pass the model creation settings to the model creation procedure through an object datatype variable v_setlst which is defined by the DBMS_DATA_MINING.SETTING_LIST datatype.

Model Testing and Evaluation

Once the HTP_CLASS_MODEL model is created, you can test and evaluate it by applying it to the test data. As shown in the Listing 6-25 PL/SQL script, to test and evaluate the newly generated HTP_CLASS_MODEL model, the DBMS_DATA_MINING.APPLY procedure applies the model to the test data in N1_TEST_DATA and generates the model scoring results stored in the HTP_APPLY_RESULT table.

Listing 6-25. Apply the Model to Test Data

```
BEGIN
  DBMS_DATA_MINING.APPLY('HTP_CLASS_MODEL','N1_TEST_DATA','CUST_ID',
  'HTP_APPLY_RESULT');
  DBMS_DATA_MINING.COMPUTE_LIFT('HTP_APPLY_RESULT','N1_TEST_DATA',
  'CUST_ID','HOME_THEATER_PACKAGE', 'N1_LIFT_TABLE', '1', 'PREDICTION',
  'PROBABILITY',100);
END;
```

The second procedure, DBMS_DATA_MINING.COMPUTE_LIFT, takes the model scoring results in the HTP_APPLY_RESULT table to compute and store the gain and lift results in the N1_LIFT_TABLE table. To review the model's lift result, a cumulative gains chart is created by querying the lift table N1_LIFT_TABLE with the SELECT statement in Listing 6-26.

Listing 6-26. Query the Lift Table

```
SELECT QUANTILE_NUMBER, GAIN_CUMULATIVE FROM N1_LIFT_TABLE;
```

To create a gain and lift chart based on this query result, QUANTILE_NUMBER in keys and GAIN_CUMULATIVE (SUM) in values are used with no groups in the notebook's visual settings, as shown in Figure 6-10.

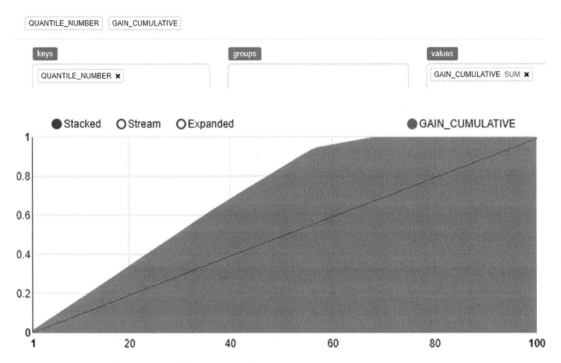

Figure 6-10. *The example's gain and lift chart*

Let's compare this chart with the standard gain and lift chart in Figure 6-2 to measure the model's effectiveness. The greater the area between the lift curve and the baseline, the better the model.

With the model accuracy being evaluated with the lift chart, you can look at the model scoring results generated from the test data by querying the HTP_APPLY_RESULT table. The query in Listing 6-27 fetches all the customers whose probability of having the home theater package (PREDICTION=1) is greater than 50%.

Listing 6-27. Fetch the Model Results

```
SELECT CUST_ID, PREDICTION PRED, ROUND(PROBABILITY,3) PROBABILITY
FROM HTP_APPLY_RESULT WHERE PREDICTION = 1 AND PROBABILITY > 0.5
```

You can join the test data table N1_TEST_DATA and the result table HTP_APPLY_RESULT and fetches all the related data for the customers who are predicted to have the home theater package (PREDICTION=1) using the query in Listing 6-28.

Listing 6-28. Join the Test Table with the Results Table

```
SELECT A.*, B.*
  FROM HTP_APPLY_RESULT A, N1_TEST_DATA B
  WHERE PREDICTION = ${PREDICTION='1','1'|'0'} AND A.CUST_ID = B.CUST_ID;
```

Model Application

During the model application, the model is applied to a new set of data. The N1_NEW_ DATA table has new customer information except for the target attribute HOME_ THEATER_PACKAGE value. You can use the model to predict the probability that each customer in the N1_ NEW_DATA will have the home theater package with this query in Listing 6-29.

Listing 6-29. Predict the Probability for New Data

```
SELECT *
FROM (SELECT CUST_ID, PREDICTION_PROBABILITY(HTP_CLASS_MODEL, '1'
      USING A.*) PROBABILITY
      FROM N1_NEW_DATA A);
```

To predict the probabilities for a given set of customers, such as those with customer IDs (102837, 103434, 103459). You just need to run a simple query, as shown in Listing 6-30.

Listing 6-30. Predict Probabilities for a Set of Customers

```
SELECT CUST_ID, ROUND(PREDICTION_PROBABILITY(HTP_CLASS_MODEL, '1'
USING A.*), 2) PROBABILITY FROM N1_NEW_DATA A
where CUST_ID in (102837, 103434, 103459);
```

Figure 6-11 shows the results from this query.

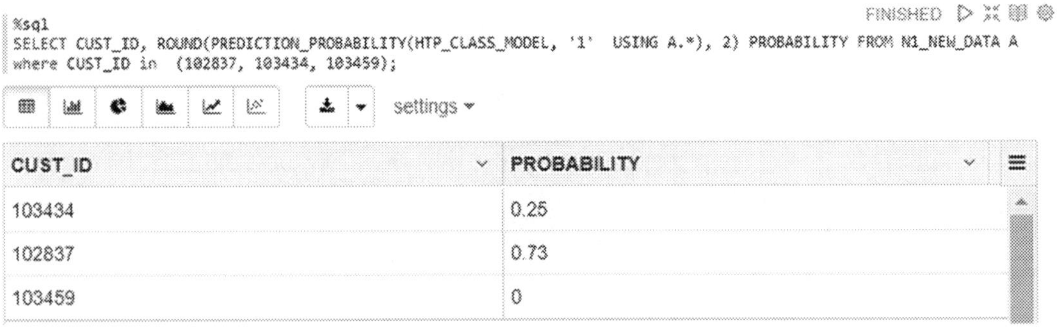

```
%sql                                                                    FINISHED ▷ ⋉ 🔳 ⚙
SELECT CUST_ID, ROUND(PREDICTION_PROBABILITY(HTP_CLASS_MODEL, '1'  USING A.*), 2) PROBABILITY FROM N1_NEW_DATA A
where CUST_ID in  (102837, 103434, 103459);
```

CUST_ID	∨	PROBABILITY	∨	≡
103434		0.25		
102837		0.73		
103459		0		

Figure 6-11. *Predict the probabilities for a given set of customers*

The same query can be joined with the N1_TEST_DATA table to predict probability for all customers, as shown in Listing 6-31.

Listing 6-31. Predict Probabilities for All New Customers

```
SELECT A.CUST_ID, A.PROBABILITY, B.EDUCATION, B.OCCUPATION,
B.HOUSEHOLD_SIZE, B.YRS_RESIDENCE, B.Y_BOX_GAMES
FROM (SELECT * FROM (SELECT CUST_ID,
    ROUND(PREDICTION_PROBABILITY(HTP_CLASS_MODEL, '1'  USING
    C.*), 2) PROBABILITY FROM N1_NEW_DATA C)) A,
    N1_TEST_DATA B
WHERE A.CUST_ID = B.CUST_ID
ORDER by A.PROBABILITY DESC;
```

Figure 6-12 shows the query and the results as displayed in OML Notebooks.

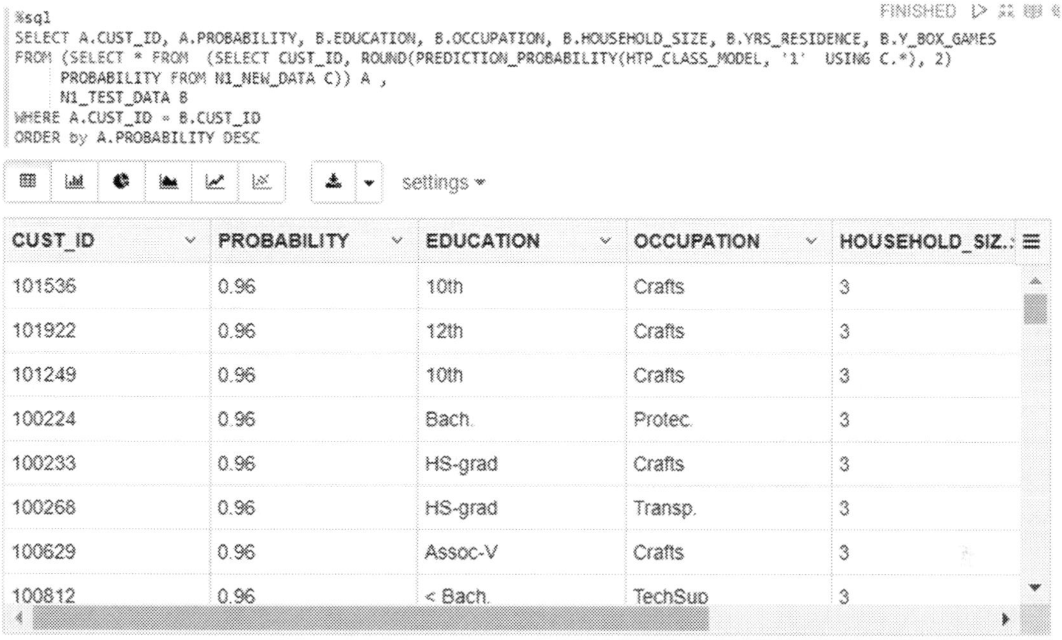

```
%sql                                                            FINISHED ▷ ⚙ ▥ ⟨
SELECT A.CUST_ID, A.PROBABILITY, B.EDUCATION, B.OCCUPATION, B.HOUSEHOLD_SIZE, B.YRS_RESIDENCE, B.Y_BOX_GAMES
FROM (SELECT * FROM  (SELECT CUST_ID, ROUND(PREDICTION_PROBABILITY(HTP_CLASS_MODEL, '1'  USING C.*), 2)
      PROBABILITY FROM N1_NEW_DATA C)) A ,
    N1_TEST_DATA B
WHERE A.CUST_ID = B.CUST_ID
ORDER by A.PROBABILITY DESC
```

CUST_ID	˅	PROBABILITY	˅	EDUCATION	˅	OCCUPATION	˅	HOUSEHOLD_SIZ.: ☰
101536		0.96		10th		Crafts		3
101922		0.96		12th		Crafts		3
101249		0.96		10th		Crafts		3
100224		0.96		Bach.		Protec.		3
100233		0.96		HS-grad		Crafts		3
100268		0.96		HS-grad		Transp.		3
100629		0.96		Assoc-V		Crafts		3
100812		0.96		< Bach.		TechSup		3

Figure 6-12. *Predict the probabilities for all new customers*

You can create a new table HTP_NEW_PREDICTION by populating it with a query like the one in Listing 6-32.

Listing 6-32. Populate a New Prediction Table with Predicted Probabilities

```
CREATE TABLE HTP_NEW_PREDICTION AS
SELECT A.CUST_ID, A.PROBABILITY, B.EDUCATION, B.OCCUPATION,
B.HOUSEHOLD_SIZE, B.YRS_RESIDENCE, B.Y_BOX_GAMES
FROM (SELECT * FROM (SELECT CUST_ID,
      ROUND(PREDICTION_PROBABILITY(HTP_CLASS_MODEL, '1' USING
      C.*), 2) PROBABILITY FROM N1_NEW_DATA C)) A ,
    N1_TEST_DATA B
where A.CUST_ID = B.CUST_ID;
```

You can then get the distribution of the probability values by running the SELECT statement in Listing 6-33.

Listing 6-33. Query the New Prediction Tables

```
select * from HTP_NEW_PREDICTION;
```

You can display the query result as a bar chart with PROBABILITY in keys and CUST_ ID COUNT in values. Figure 6-13 shows the probability bar chart.

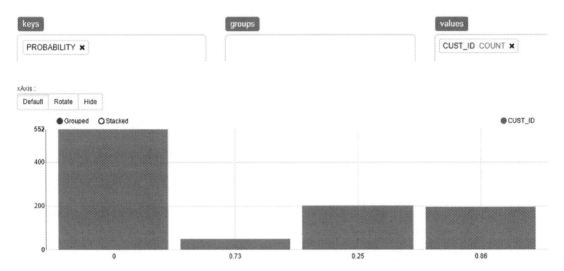

Figure 6-13. *The customer-count bar chart by probability*

Summary

This chapter focused on building, evaluating, and applying machine learning models in the Oracle Machine Learning environment with Oracle Autonomous Database. It introduced the major phases of the machine learning process flow and two types of machine learning functions: supervised machine learning and unsupervised machine learning and the related algorithms supported by Oracle Machine Learning.

We discussed how to leverage Oracle Machine Learning for SQL (OML4 SQL) in different machine learning process flow phases, such as data preparation and data transformation, model creation, model evaluation, and model scoring and application. We discussed some of the key OML4SQL API packages, such as DBMS_DATA_MINING, and DBMS_DATA_MING_TRANSFORM, and their procedures, including CREATE_ MODEL, CREATE_MODEL2, APPLY, and COMPUTE_LIFT. We also discussed the SQL functions for model scoring, such as PREDICTION and PREDICTION_PROBABILITY.

The last section of the chapter used a step-by-step example of how to build, evaluate and apply machine learning models with OML4SQL API and OML Notebooks in ADW using the sample data available in the ADW database sample schema. You also can follow this example to go through this example in an ADW OML Notebooks environment once you sign up for an Oracle Cloud account and create an ADW database instance.

CHAPTER 7

Oracle Analytics Cloud

This chapter focuses on Oracle Analytics Cloud. Oracle has three product lines for analytics: Oracle Analytics Cloud (OAC), Oracle Analytics Server, and Oracle Analytics for Applications. Oracle Analytics Desktop is available for standalone data exploration and visualization on your own computer. Oracle's vision for analytics is that it should be augmented, integrated, and collaborative. Data preparation and visualization are very important steps in the machine learning process, and OAC is a very good tool for that, especially because of its augmented capabilities. Augmented analytics is powered by machine learning. OAC supports integration very well by allowing users to add data from many different data sources, including Oracle SaaS applications.

OAC also allows you to create machine learning models (ML models), register and use ML models created in Oracle Database, and create scenarios for a new data set based on an ML model. It also offers many ways to easily and efficiently prepare the data and store the prepared data, for example as a table in Oracle Autonomous Datawarehouse (ADW) for further ML processing. OAC supports collaboration in many ways, including fine-tuned privileges and the possibility to publish the analysis on different platforms.

Currently, there are two editions of OAC: Oracle Analytics Cloud Professional Edition and Oracle Analytics Cloud Enterprise Edition. The Professional Edition includes all functionalities for data analytics. The Enterprise Editions includes even more, such as data modeling, pixel-prefect reporting, and different migration possibilities from other products and environments.

OAC is not part of the free offerings, but it can be used with the 30-day free trial. The desktop edition (Oracle Analytics Desktop) can be used for testing and training for free, but there is a license fee if you use it for production. Oracle Analytics Server brings the OAC platform functionalities to an on-premise deployment. Oracle Analytics for Applications provides analytics for Oracle Cloud applications, powered by Autonomous Data Warehouse and Oracle Analytics.

© Heli Helskyaho, Jean Yu, Kai Yu 2021
H. Helskyaho et al., *Machine Learning for Oracle Database Professionals*,
https://doi.org/10.1007/978-1-4842-7032-5_7

In Figure 7-1, you can see the main navigator menu of OAC. Home takes you to the main page that has all the content in OAC and a search functionality. Catalog lists all the folders and their contents. Data has all the data sets, connections, data flows, sequences, and data replication definitions. Machine Learning includes all the machine learning models created with OAC or registered from an Oracle Database. Jobs lists all jobs and schedules defined in OAC. Console lists tools for visualizations and sharing, configuration, and administration. Visualization and sharing tools include maps, extensions, and social. Configuration and administration tools include search index, safe domains, users and roles, snapshots, connections, virus scanner, session and query cache, issue SQL, monitor deliveries, mail settings, system settings, and remote data connectivity. Academy takes you to a page full of learning possibilities about OAC. Starting with getting started to advanced topics like setup and manage or how to use pixel-perfect reports.

In OAC, a Project is a collection of data preparations, visualizations, and narrating and it can be shared in different format with other users.

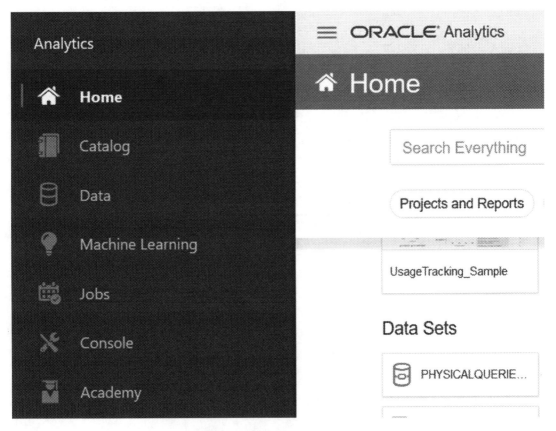

Figure 7-1. *Main navigator menu for OAC*

Data Preparation

As with machine learning, data analytics start with the business problem and define what you want to achieve. Then you need the data to support the task, the business problem.

Using the Create button (see Figure 7-2), you can create projects, data sets, data flows, sequences, connections, data replications, and replication connections. A *project* is a collection that holds your data sets, visualizations, and all the elements you want to connect, called a *data visualization project*. You can create a project from the data set you are preparing by selecting Create Project on the canvas. You can also create a project by opening a data set and clicking Save. You can create a project using the Create button on the Home screen and selecting Data Set.

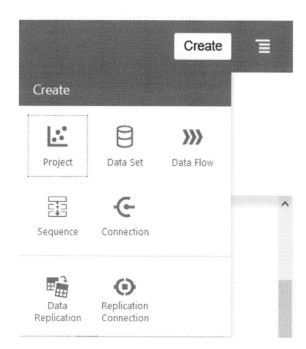

Figure 7-2. *Create*

You can create a data set by selecting Create ➤ Data Set (see Figure 7-2). Or you can open an existing data set by selecting either Home or Data from the main menu and finding the data set you are looking for. The data set can be created from a file or a connection to an external data source. If you want to use a file, you can just drag and drop the file. If you want to use a connection, you can use data from existing connections or create a new connection. When creating the connection, you start it by selecting one of the many connection types. Figure 7-3 shows the available connection types at the time this book was written. You can also use ODBC and JDBC to connect other databases that support those connectors.

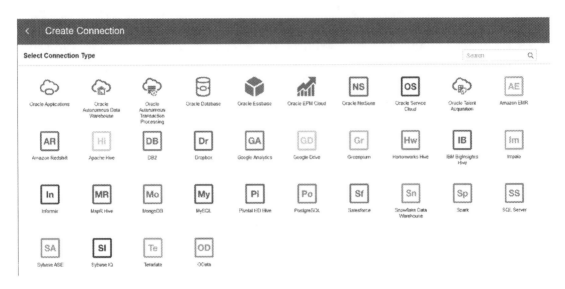

Figure 7-3. *Create Connection, Select Connection Type*

When you open a data set, column-level profiling is automatically run against a sample of the data set. After profiling the data set, OAC recommends some transformations and enrichments to the data. You can see these recommendations in the Prepare tab under Recommendations. Check the recommendations, and either accept or ignore them. The recommendations can be augmented data enrichment, such as sensitive data detection, or geographic or demographic enrichments. They can be transformations, like dividing the name column into first name and last name columns or removing the social security number from the data set.

You can also add your own transformations or enrichments using the column's Options menu. The list of available menu options for a column depends on the type of data in that column. In the middle of the screen, you can see a sample of the data. When you transform or enrich data, either using the recommendations or manual steps, a step is automatically added to the Preparation Script pane and run to the sample set of data. You can add, edit, or remove action steps in the Preparation Script. After checking the recommendation, check each column and see if there is something you should do with it. Maybe change the column name, the data type, the treat as (attribute/measure), the formatting, or maybe create groups or binning, for example. A blue dot next to a transformation step in the Preparation Script indicates that the script has not been run for the whole data set. To run the script to the whole data set, click Apply Script.

You can prepare several data sets at the same time by selecting each of them. If you click the Data Diagram tab on the bottom of the screen, you can see how OAC thinks the data sets a link to each other. If it is correct, you do not need to do anything. If it is not, create an accurate link for the data sets. The data transformations and enrichments that you apply to a data set affect all projects and data flows where that data set is used. If you do not want that to happen, you can use the Duplicate Data Set functionality to keep the data set separate from other data processing. Saving the data transformations and enrichments creates a project.

Use a data flow if you have several data sets to combine, or if you want to save the result in a new data set. You can prepare the data sets with the Prepare functionalities, and then add those prepared data sets to a data flow. A data flow creates sequential transformations and result in a new data set for a project to use without changing the original data sets. Using the Branch operation, you can create several outputs from a data flow. In a data flow, you can use filters to limit the amount of data, add columns to include more data, add new data sets to extend dimensionality or facts, select columns to focus on relevant data attributes, aggregate data, rename columns, and much more.

A data flow can be run manually or scheduled. The schedule option is only available in the cloud. You can create a data flow by selecting Create ➤ Data Flow, as shown in Figure 7-2. You can see the Data Flow Steps menu to the left of the Data Flow canvas in Figure 7-4.

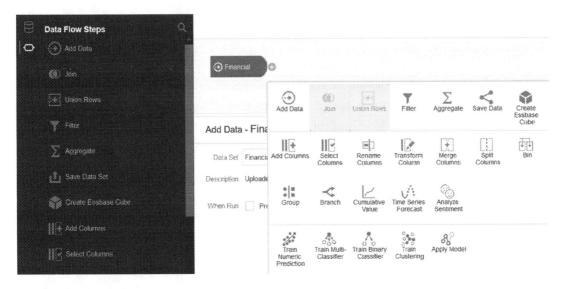

Figure 7-4. *Data and data flow*

These are the possible steps you can add to a data flow. You start by selecting Add Data and selecting the data set(s) needed. If you add several data sets, the tool automatically creates a join for the data to find columns for joining. Then for each data set, define the steps you want to perform. For example, you can add, select, rename, transform, merge, or split columns. You can bin or group data and use a branch to separate two different outcomes. You can filter or aggregate data and so on. The last step in a data flow is always Save Data. The data can be saved to OAC or to an external target database that you can connect to. This makes OAC an excellent tool for preparing data. In Oracle Machine Learning (OML) processing, data from different sources can be combined and prepared using OAC and stored to ADW to be used with OML, as shown in Figure 7-5.

Save Data Set

		Columns	
Data Set	MISSING_PRODUCTS	Name	Database Name
Description		Product	PRODUCT
		Brand	BRAND
Save data to	Database Connection ▾	LOB	LOB
Connection	☁ ADW11	Product Type	PRODUCT_TYPE
Table	MISSING_PRODUCTS_TO_ANALYZE	Sales Rep Name	SALES_REP_NAME
When run	Replace existing data ▾	Lastname	LASTNAME
When Run	☐ Prompt to specify Data Set		

Figure 7-5. *Saving the result of a data flow to a table in ADW*

A *sequence* is a set of data flows for executing multiple data flows. Sequences can be run manually or scheduled. The scheduling option is only available in the cloud. You can create a sequence by adding several data flows in a chain and executing it.

You can also start the preparation process by creating a project. Simply click the Create button and select Project (see Figure 7-2). When you create a project, the next step is to choose the data sets for it. You can select from the list of available data sets or create a new data set.

When a project has been created and the data sets selected, you can start working on it. If you have selected several data sets, the tool tries to find data to join them, or you can manually do it. The first thing to do is prepare the data if you have not done so already. In the upper-right corner, you find three tabs: Prepare, Visualize, and Narrate. For data preparation, select Prepare. On the left-hand side of the canvas, you find two options for Prepare: Data and Preparation Script. If you select Data, you see the selected data sets, prepare them, and add new data sets to the project. If you select Preparation Script, you see the steps to prepare the data. You can add steps to the script by accepting the steps suggested by OAC or by adding your own steps.

Data Visualization and Narrate

If you select Visualize from the options on the right side of the canvas, you see three options for working on a project: Data, Visualization, and Analytics.

You can add more data sets to the project by selecting the + icon beside the word Data and selecting Add Data Set. You can work on the selected data sets by selecting the Data button. You can also create scenarios (see Figure 7-6) or add calculations (see Figure 7-8).

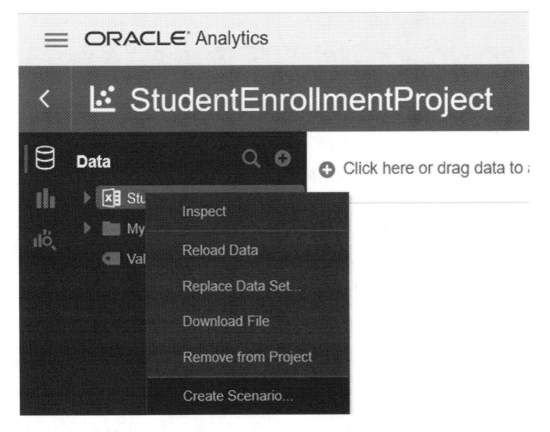

Figure 7-6. *Adding a scenario*

To create scenarios, you select a model from the list of pre-created scenarios. Then you map the data set for the data in the model. The data set and the model calculate the scenario. The scenario is shown in the same navigator as the data sets (see Figure 7-7) and can be used for visualization.

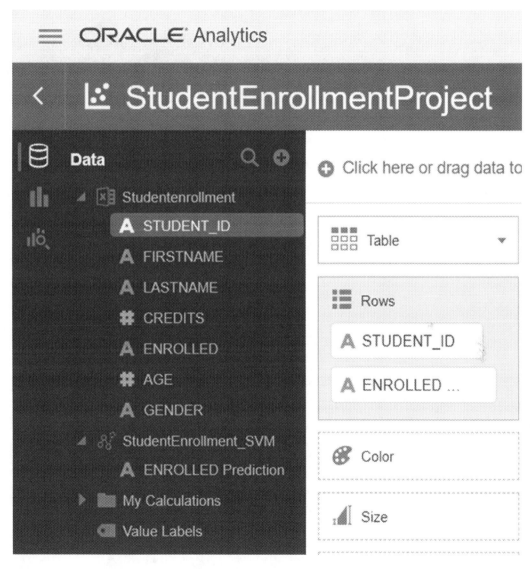

Figure 7-7. *Using an ML model to predict*

To add a new calculation, select Add Calculation (see Figure 7-8). Give the calculation a name, select the mathematical function from the list, and then fill in the information needed. In Figure 7-8, we chose Aggregate. By double-clicking the word *measure*, you get a list of possible measures from the data set. By double-clicking the word *dimension*, you get a list of dimensions.

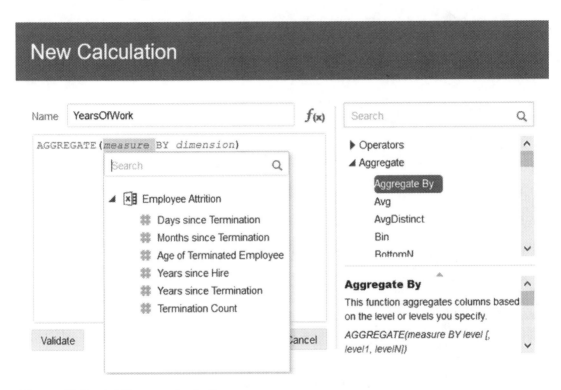

Figure 7-8. *Adding a calculation*

To create visualizations, you can select the attributes (hold the Ctrl key and select all the attributes needed) and then drag and drop them into the canvas. Or, you can right-click and select Best Visualization to let the tool decide which chart is best suited to the selected data. Or, you can select Visualizations and choose from the options shown in Figure 7-9.

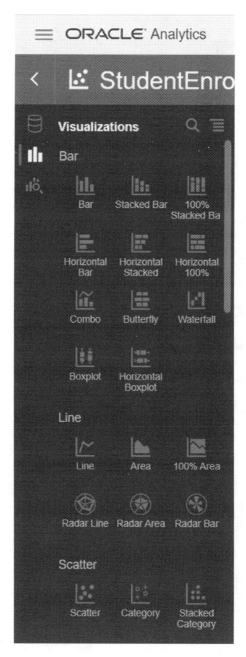

Figure 7-9. *Visualization options*

If you want to get a head start to analyses, you might want to try the Explain feature. The Explain feature (see Figure 7-10) automatically explains the attribute chosen.

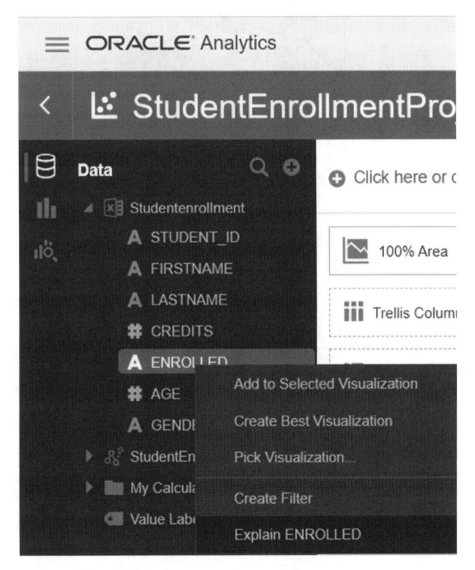

Figure 7-10. *Explain an attribute*

The Explain feature functionality explains the attribute, including its key drivers and segments, and any anomalies in the data set regarding it. All of these are explained in charts and text. If you like any or some of them, you can use them as a basis for your analysis by selecting them for a canvas, as shown in Figure 7-11.

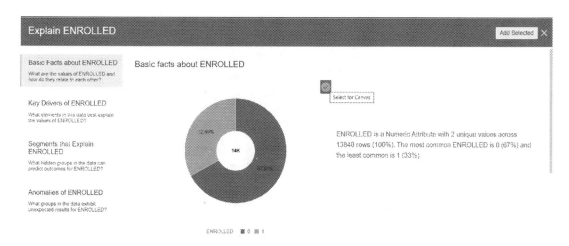

Figure 7-11. *Select for canvas from the Explain report*

There are at least two good reasons to start with the Explain feature: avoid cognitive bias and start the task. Explain gives you a good understanding of the data.

You can add, edit, deleted, copy, and duplicate visualization. You can add and remove data, filters, charts, canvasses, and so forth. You can fine-tune the charts with the chart properties in the bottom-left corner of the canvas. You can create your own extensions (plug-ins), or you can use existing extensions available at www.oracle.com/ business-analytics/data-visualization/extensions.html. You can add those plug-ins to your OAC in the Console under Extensions ➤ Upload. And you can select Language Narrative as a visualization type from the list to let OAC automatically create a narrative of the visualization. This is an easy way to rapidly generate a foundation for a report.

If you click the Analytics tool, you can add information about clusters or outliers or add a reference line, trend line, or a forecast to the chart (see Figure 7-12).

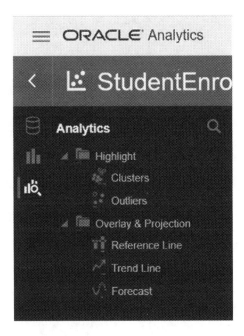

Figure 7-12. *Analytics options*

You can create reports or presentations from the analysis and visualizations you have created using the Narrate tab. You can choose canvases from the visualizations, add descriptions to them, hide charts, add notes, and so forth. And you can present the narration from OAC. Or you can export the project to PowerPoint (.pptx), Acrobat (.pdf), image (.png), or OAC package (.dva) formats, or simply print it out.

Machine Learning in Oracle Analytics Cloud

You already saw several examples of machine learning being used in OAC. When the data was uploaded to OAC, the data sample finds recommendations for transformations and enrichments. The Explain functionality, automatic language narrative, and some other features are examples of machine learning to make the tool easier and more efficient. But you can also use machine learning in other ways.

You can use data flows to create predictive models. Use the Data Flow editor to first create and train a model on training data (see Figure 7-13), then apply it to the data sets with known input and output to evaluate it. You can iterate the training process until the model reaches the quality you want. Finally, use the finished model to score/predict the unknown outputs on new data and save the predicted data set. You can also apply the predictive model to a project by adding a scenario.

200

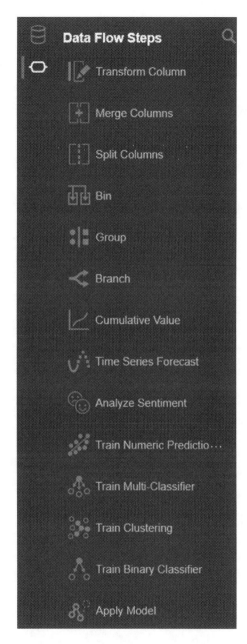

Figure 7-13. *Machine Learning Model training in OAC*

To see available machine learning models, select Machine Leaning from the main navigator menu (see Figure 7-1). If you select a machine learning model from the available models, you can investigate it, mark it as your favorite, or delete it. If you cannot find the model from the list, you can register the model. Registering allows you to use an existing machine learning model created in Oracle Database. Select Register ML Model, as shown in Figure 7-14.

Figure 7-14. *Registering an existing machine learning model*

To register an ML model, you select a connection to the Oracle Database where that model exists (see Figure 7-15). This can be to an on-premise database or Oracle Autonomous Database. If there is no connection available to the database where the ML model locates, you need to create a new connection with the Create Connection button.

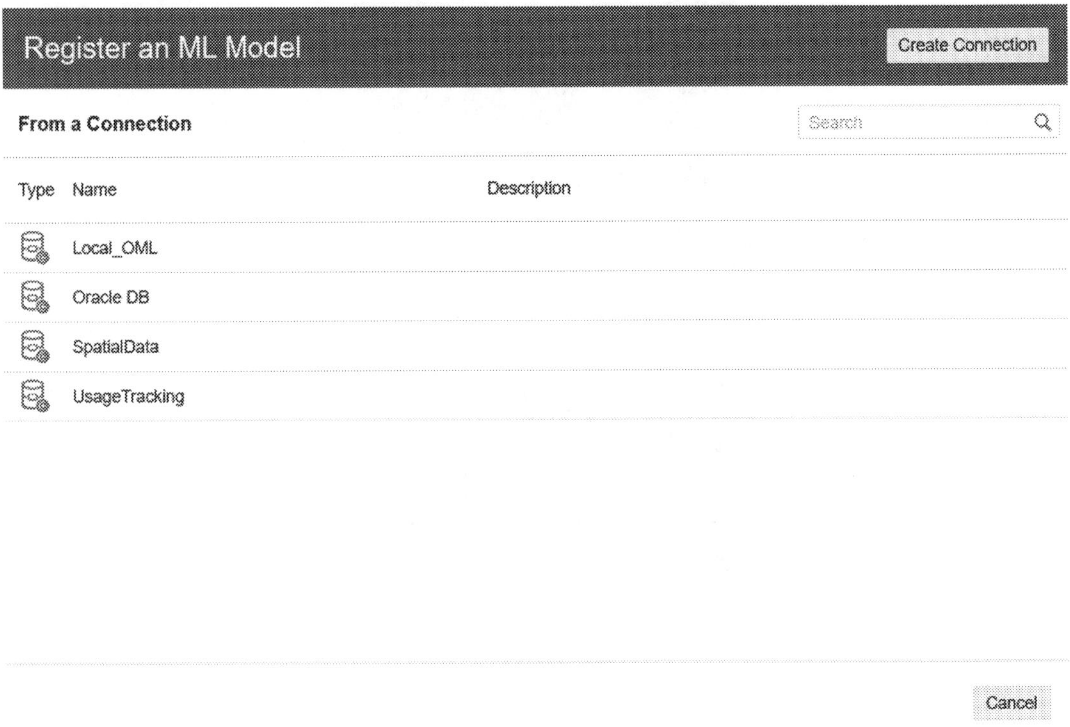

Figure 7-15. *Select a connection or create a new connection to register an ML model for OAC*

When you have connected to the database, you see a list of available ML models. Select the model and click Register. Now the model is usable in OAC.

You can also use machine learning for the search functionality on the Home screen of OAC. You can click the microphone icon and verbally ask your analytical question. OAC checks the data, should find the data to answer the question, and automatically creates the charts for the analysis. OAC uses a lot of machine learning to help users analyze the data and build reports and predictions based on it. Each release of the tool always brings more of these interesting functions. This tool is worth taking a deeper look at.

Summary

This chapter discussed Oracle Analytics Cloud, one of the products Oracle offers for analytic needs. OAC is a good tool for analytics and an excellent tool for preparing the data for machine learning. Oracle's vision for analytics is that it should be augmented, integrated, and collaborative. This is the reason for OAC to be an excellent choice also for machine learning. You can easily combine data from different sources and use augmented functionalities to prepare it. After you are done preparing the data, you can store it in a database for further use. OAC's Explain feature makes it easier to understand the data. You can use machine learning models from Oracle Database with OAC or build simple machine learning models. OAC is a good option for the machine learning process since it is well integrated with other Oracle products and other possible data sources, and it has strongly augmented functionalities.

CHAPTER 8

Delivery and Automation Pipeline in Machine Learning

Growing volumes of available data, cheaper and powerful computational processing, and more affordable data storage continue to accelerate and fuel the integration of business applications and machine learning (ML) models. ML models power more and more applications in production. The following are some well-known model-driven applications.

- Online recommendation systems used by Netflix and Amazon

- Trendy self-driving cars from Tesla and Alphabet

- Fraud detection solutions for banks and e-commerce merchants

- Machine learning algorithms for medical diagnosis and prediction

- Sentiment analyzer on social media to provide digital consumer insights

This chapter explores the issues and challenges of machine learning development and deployment. It covers some of the key requirements and design considerations behind ML deployment solutions. Finally, it dives into specific solution components, best practices, and tooling.

© Heli Helskyaho, Jean Yu, Kai Yu 2021
H. Helskyaho et al., *Machine Learning for Oracle Database Professionals*,
https://doi.org/10.1007/978-1-4842-7032-5_8

ML Development Challenges

Like any software solution, model-driven applications are measured using common software quality characteristics, such as functionality, reliability, maintainability, scalability, and performance. ML applications must address all the quality issues of traditional software plus an additional set of ML-specific requirements. Therefore, ML application development is harder than traditional software development.

Classical Software Engineering vs. Machine Learning

In classical software engineering settings, the desired product behavior can be sufficiently expressed in software logic alone. Once the coded logic is thoroughly tested in user scenarios with mock data, the code is expected to continue to work with the real data in the production environment.

On the contrary, an ML application learns to perform a task without being specifically programmed. Its quality depends on the input data and tuning parameters. When a model is trained, a mapping function is established between the input (i.e., independent variables) to the output (i.e., target variables). This mapping learned from historical data may not always be valid in the future on new data if the relationships between input and output data have changed. This change can have detrimental effects on the dependent applications or services because the model predictions become less accurate as time passes. An ML application must be fed new data to keep working.

Model Drift

The concept of model drift means that, over time, changing external data results in poor and degrading predictive performance in predictive models. Depending on the cause of the drift, model drift can be categorized into two types: concept drift and data drift.

Concept drift refers to the change in the statistical properties of the target variable change over time. Email spam filtering application exhibits frequent concept drift due to the evolving nature of spam and legitimate emails.

While concept drift is about the target variable, data drift occurs when properties of independent variables change. For instance, a data set contains information on default payments, demographic factors, credit data, history of payment, and bill statements of credit card clients. A machine learning model can be trained to predict the likelihood of credit card default for customers in the bank. In this case, it is not the definition of the target variable (default payment) that changes but the values of independent variables (demographic factors, credit data, history of payment, and bill statements) that define the target.

Note   In machine learning, features or attributes are independent variables input for the machine learning algorithm to analyze. A target (the output) is the dependent variable about which you want to gain a deeper understanding. In the context of a supervised classification model, the term class label refers to a unique value of the target.

ML Deployment Challenges

As data scientists and software engineers continue to accumulate experience with live ML systems, you can gain a greater understanding of the unique challenges, how they happen, their implications, and how to overcome the challenges.

Before being deployed, models are pre-trained and typically exported to one of the file formats (e.g., pickle, ONNX, PMML). Without model drift, the Model Serving code can load these pre-trained model packages and make predictions to serve the downstream application.

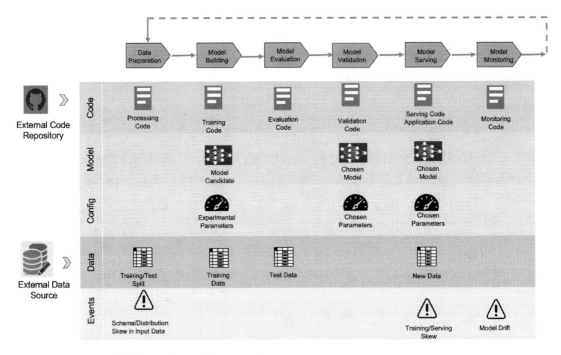

Figure 8-1. *ML life cycle and key artifacts*

For models that deal with live data, model training does not end at model development. Once operational, models need to be monitored, and frequent model retraining may be required to keep up with changing circumstances. A retraining pipeline with model monitoring needs to be added to feed the model serving module, resulting in the life cycle pattern in Figure 8-1.

ML Life Cycle

The ultimate goal of deploying and maintaining live ML-driven applications is to maintain optimal model performance metrics (such as loss, accuracy). This requires constant experimentation/validation to improve metrics due to changes in live data. To make this task more challenging, the improvement needs to be automated.

As described in Figure 8-1, each iteration of the ML life cycle is associated with a specific version of the source code (including versioned dependent libraries) pushed from an external repository. Also, the iteration accesses input data from an external data source which is also versioned.

Operations in all phases are required to be orchestrated, so the life cycle can be fully automated. Let's look at the additional artifacts generated during processing in each phase.

- **Data preparation**: Data processing code first validates the raw data and generates events if schema or distribution skew detected. Otherwise, the data is pre-processed and split into a training set and a test set.

- **Model building**: A candidate model is created with experimental parameters. This model is trained using the training set.

- **Model evaluation**: The candidate model is evaluated, and performance metrics are produced.

- **Model validation**: The candidate model is validated/compared with the current live version. The validation code chooses a version that better meets the performance requirement.

- **Model serving**: The model serving code promotes the chosen model to production to serve predictions using new data.

- **Model monitoring**: In this phase, schema or distribution skews (in features and predictions) are detected by comparing consecutive spans of historical data sets. For instance, if data ingestion happens monthly, you compare between consecutive months of training data. Model drift events are logged if notable drift is detected, which triggers model retraining by starting a new iteration of the life cycle.

Scaling Challenges

Data scientists and data engineers focus on several key areas when it comes to scaling machine learning solutions.

Model Training

During training, a series of mathematical computations are applied to massive amounts of data repeatedly. Training such a model to reach a decent accuracy can take a long time (days or weeks). This problem is multiplied by the number of models to be trained.

Model Inference

Model inference service has to handle many inputs or users and meet latency and throughput requirements. For accelerating deep learning inference in computer vision, optimization is required in both hardware and application levels.

Input Data Processing

Throughout a typical ML life cycle, a massive amount of data is fed to the system. The memory representation and the way algorithms access data can have a significant impact on scalability. Furthermore, to keep up with the data consumption rate of multiple accelerated hardware devices, the data processing speed and throughput must be scaled up.

Key Requirements

Based on the analysis, the key requirements for supporting live ML models are as follows.

- Automation of the entire ML life cycle (with the possibility of human intervention/approval when required).

- Tracking and managing ML models and artifacts such as data, code, model packages, configuration, dependencies, and performance metrics are critical for drift detection, model reproducibility, auditability, debugging, and explainability.

- Model performance (such as accuracy) is the key quality indication of any ML model.

- ML operational requirements include ML API endpoints response time, Memory/CPU usage when performing prediction, and disk utilization (if applicable).

Design Considerations and Solutions

To overcome challenges and meet the set of requirements, let's discuss common design considerations and recent advances in solutions and tooling.

Automating Data Science Steps

Typically, productizing any ML model requires automating the life-cycle processing, which translates to the following specific data science steps.

- **Data ingestion/pre-processing**: Collect and process raw data from incoming data sources.

- **Data validation**: Verify the input data schema and statistical properties are as expected by the model.

- **Data preparation**: Clean, normalize, and split the data into training/test sets.

- **Feature engineering**: Transform raw data to construct additional features that better represent the underlying problem to the model.

- **Feature selection**: Remove irrelevant features and select those features which contribute most to the target variable. The output of this step is the feature splits in the format expected by the training algorithm.

- **Model training**: Apply various ML algorithms to the prepared feature set to train ML models. This includes hyperparameter tuning to get the best performing ML model. The output of this step is a trained model.

- **Model evaluation**: Assess the quality of the trained model using a set of metrics.

- **Model validation**: Determine whether the model is adequate for deployment based on model evaluation and monitoring metrics.

- **Model serving**: The ML inference program uses a trained model to generate predictions.

- **Model monitoring**: Monitor to detect training/serving skew and model drift. If the drift is significant, the model must be retrained and refreshed with more recent data.

These manual data science steps must be reflected in an automated pipeline before the pipeline can be deployed in any production environment. MLOps is an important and relatively new concept in the context of ML automation. Let's introduce MLOps and then dive into key solutions that are enabled within the MLOps framework.

Automated ML Pipeline: MLOps

MLOps (a compound of *machine learning* and *operations*) is a set of practices that combines the practices between data scientists and operations professionals to automate and manage the production ML (or deep learning) life cycle. MLOps is a CI/CD framework for large-scale machine learning productization.

Note   *Continuous integration* (CI) is a software development best practice for merging all developers' working copies to a shared repository frequently, preferably several times a day. Each merge can then be verified by an automated build and automated acceptance tests.

Continuous delivery (CD) is the practice of keeping your codebase deployable at any point. Beyond making sure your application passes automated tests, it must have all the configuration, dependencies, and packaging necessary to push it into production.

DevOps is a set of practices that automate and integrate the processes between classical software development and operations teams. DevOps combines cultural philosophies, practices, and tools to increase an organization's ability to deliver high-velocity applications and services.

The operational side of the MLOps process is similar to DevOps. MLOps can reuse the classic DevOps toolchain components to handle source code repository, build/test/packaging/deployment automation, and integration test, such as the following.

- Git: A source code repository

- Jenkins: An automation facility

- Kubeflow and Airflow: Orchestrators

- Travis CI: A continuous integration service

- Argo CD: A continuous delivery tool

- Kubernetes: A container-orchestration system for automating application deployment

Figure 8-2 shows a model-driven application code contains an ML pipeline (step 1). Once the code is committed and merged in the code repository (Git), DevOps' continuous integration tool automatically starts the integration and testing (step 2).

The continuous delivery tool makes sure the ML pipeline is packaged with all necessary configurations/dependent libraries and ready to be deployed to production (step 3).

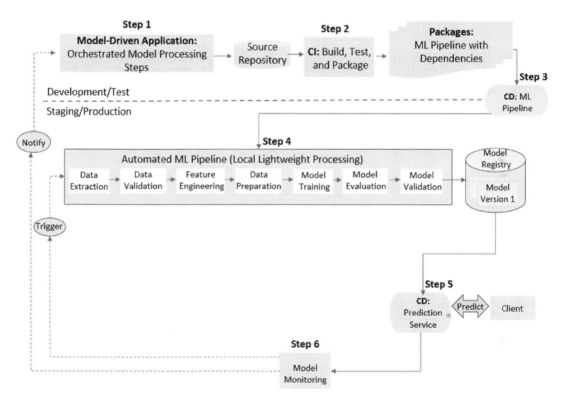

Figure 8-2. *ML pipeline automation*

However, production ML presents additional CI/CD and orchestration challenges and nuances due to ML-specific interdependencies. MLOps is an extension of DevOps with additional infrastructure and tooling.

- **Special hardware**: Special segmentation and orchestration of data and algorithm required to take advantage of accelerated hardware (such as Graphics Processing Units (GPUs) and Tensor Processing Units (TPU)) and meet scalability requirements.

- **Complex computational graphs**: Orchestration can involve complex runtime computational graphs. ML applications can have multiple pipelines executing in parallel.

  - A/B testing

  - Ensembles

  - Multi-armed bandits

- **Reproducibility**: Due to the experimental nature of model training, it is important to maintain repeatability for pipelines and components. The ML pipelines can be reused, adjusted, resumed using the information in the model registry. For instance, a pipeline can skip some steps or resume an intermediate step (step 4).

- **Continuous model delivery**: Once the model registry decides to raise a model from staging to production, the model is automatically made ready to be consumed by the Prediction Service and the downstream application/client (step 5). This step enables the "continuous delivery" of new model versions within an ML pipeline, which overrides the regular DevOps process (step 3). Additional ML-specific interdependencies such as model registry, model refresh management, model validation, data validation, or drift detection.

- **Regulatory compliance**: Additional explainability and interpretability modules must articulate model intricacies to address various compliance regulators.

Model Registry for Tracking

The model registry component is a centralized data store that supports the management of the full life cycle of ML models. It provides model versioning, production/stage/archived transitions, tracking for data dependency, code dependency, environment, and configuration for every active pipeline instance.

As displayed in the example in Figure 8-3, the automated ML pipeline produced multiple versions of the trained model: v1, v2, and v3. Using the model registry information, the model validation/monitoring modules decided to deploy v1 to production. V1 was later archived and replaced by the more optimal v2. This mechanism allows the continuous model refresh/rollback triggered by model monitoring.

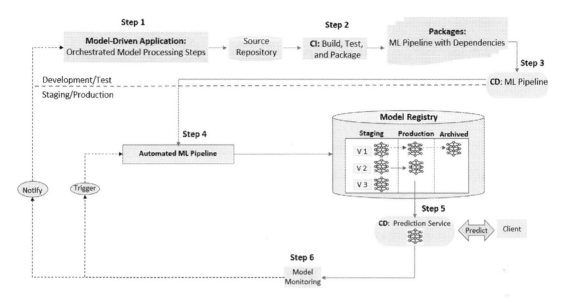

Figure 8-3. *Model registry example*

The following items are typically tracked and saved in the model registry.

- The selected version of the trained model

- The version of input data (the version to identify the raw input data)

- The version of code (versions of all source code—model training, application, pipeline processing, etc.—as tracked in the external source code repository)

- Dependencies (linkage to dependent libraries, database connections, etc.)

- Environment configuration (parameters to control ML pipeline components, such as data processing, feature selection, the threshold value used by data validation or model monitoring)

- Hyperparameters (a parameter whose value controls the ML learning process; e.g., learning rate, batch size, size of a neural network, number of epochs, number of layers, number of filters, and optimizer name)

- Trained model packages(the resulting package from model training. Typical formats are pickle, ONNX, and PMML)

- Model scoring results (the predictions made by the model)

- Features for training and testing

- Feature importance (the score assigned to input features based on how useful they are at predicting a target variable)

- Model performance metrics

  - Training time

  - Accuracy

  - Loss

  - Confusion matrix (see Figure 8-4)

  - Accuracy: $(TP+TN)/(TP+FP+TN+FN)$

  - Precision: $TP/(TP+FP)$

  - Recall: $TP/(TP+FN)$

TP, TN, FP, and FN represent True Positive, True Negative, False Positive, and False Negative. Figure 8-4 describes how TP, TN, FP, and FN are defined in a binary classifier (a classifier with two classes: True and False).

- TP is the number of cases that were predicted to be true, and the actual result is True.

- TN is the number of cases that were predicted to be True, but the actual result is False.

- FP is the number of cases that were predicted to be False, but the actual result is True.

- FN is the number of cases that were predicted to be False, and the actual result is False.

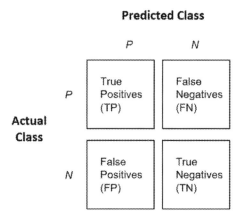

Figure 8-4. *Model evaluation: confusion matrix from binary classification*

By storing the information in the model registry, it is possible to reproduce any model training. You can also monitor a series of training instances to detect any regression and choose to redeploy or even roll back all or parts of the pipeline in production.

Data Validation

As the rapid application of ML models to key areas of our daily life, it is important to explain the reasoning behind the model predictions. However, there is no easy way to effectively examine the internal structure of a trained model for correctness to justify unintended model behavior.

By comparison, validating the data is much more straightforward. The input data is the starting point of the ML pipeline, so data validation should be the first line of defense to flag problems before the entire pipeline is activated with poor data.

Validation is performed before model training to detect data drift by looking at a series of consecutive data sets to ensure the following.

- The current data set conforms to the same schema as the previous data set.

- The feature values of the current data set are similar to those of the previous data set.

- The feature value distributions of the current data set are similar to those of the previous data set.

- Other model-specific patterns are unchanged. For example, in a Natural Language Processing (NLP) scenario, a sudden change in the number of words and the list of words not seen in the previous data set.

Data validation is also performed before the model serving to ensure the following.

- The serving data set and the training data set conform to the same schema.

- The feature values that the model trains on are similar to the feature values that it sees at the serving time.

- The feature value distributions for the training data set are similar to those of the serving data set.

- Other model-specific patterns are unchanged. For example, in an NLP scenario, a sudden change in the number of words and the list of words in the serving data not seen in the training data.

Pipeline Abstraction

The ML pipeline consists of various processing steps that are more or less the same regardless of the targeted business problems. Pipeline abstraction is a popular practice seeking to modularize the common functionalities and share them as services across all pipelines. The purpose is to reduce inconsistency in code, improve pipeline quality and maintainability, and increase the overall productivity of ML systems. The following is a list of abstract service examples.

- Data transformation

- Data normalization

- Model invocation

- Algorithm selection

- Term frequency–Inverse document frequency (TF-IDF) weighting

Automatic Machine Learning (AutoML)

While the pipeline abstraction aims to deduplicate and optimize the code structure, AutoML seeks to automate the many cumbersome and repetitive steps or groups of steps in the pipeline during runtime.

With AutoML, people with limited machine learning expertise can take advantage of the state-of-the-art modules. AutoML also frees model developers from the burden of repetitive and time-consuming tasks.

Moreover, AutoML leverages neural architecture search and transfer learning to select the combination of algorithms and parameters that produce the optimal model performance automatically. The following are some of the steps in the ML pipeline that AutoML has targeted.

- Data pre-processing

- Data partitioning

- Feature extraction

- Feature selection

- Algorithm selection

- Training

- Tuning

- Ensembling

- Deployment

- Monitoring

Model Monitoring

How do you know your models in production are behaving as you expect them to? What are the impacts on your model and the downstream application when your training data becomes stale? How does your ML pipeline handle corrupted live data or training/serving skew?

To begin addressing these complex issues, metrics and artifacts representing the raw measurements of performance or behavior are observed and collected throughout your systems. For each pipeline execution instance, model packages, metrics, and other artifacts are persisted in a model registry. Model registry is covered in more detail later in this chapter.

Using the information in the model registry, the model monitoring component can reproduce the pipeline execution, conduct a detailed comparison of the current and historical instances to detect data deviation, model drift, dependency changes, and conduct production A/B testing.

Model Monitoring Implementation

Model monitoring is performed between instances of the ML pipeline. The goal is to detect notable shifts/changes in predictions, features, and metrics.

- Schema of incoming raw data

- Distribution of incoming raw data

- Model prediction distribution for regression algorithms or prediction frequencies for classification algorithms

- Model feature distribution for numerical features or input frequencies for categorical features

- Missing value checks

- Model performance metrics (loss, accuracy, etc.)

Common metrics to calculate shift

- Median and mean values over a given timeframe

- Min/Max values

- Standard deviation over a given timeframe

If the monitoring component detects a significant drift, it triggers the restart of the ML pipeline with more recent input data. The resulting model instances are placed in staging. The model validation module then decides to refresh the model in production.

In dependency change, the monitoring module notifies the data scientist to request human intervention/approval.

Scaling Solutions

ML applications require specific hardware configurations and scalability considerations. For example, training neural networks can require powerful GPUs. Training common machine learning models can require clusters of CPUs. To take advantage of the accelerated hardware, the ML model and input data need to be partitioned, sized, configured, and mapped a cluster of computational nodes (running ML frameworks such as Spark, scikit-learn, and PyTorch).

ML Accelerators for Large Scale Model Training and Inference

Many machine learning accelerators are released for various deep learning models that support object detection, natural language processing, and assisted driving. The capabilities of the ML accelerators depend on the types of neural networks. There is also the distinction between neural network training and inference. The purpose of this section is to simply introduce a few currently well-adopted accelerators in production environments.

- GPUs (graphics processing units) contain hundreds of embedded ALUs that handle parallelized computations.

- Application-specific integrated circuits (ASICs) such as Google's TPUs are matrix processors designed for deep learning.

- Intel's Habana Goya and Gaudi chips: Gaudi for training and Goya for inference.

Distributed Machine Learning for Model Training

Distributed machine learning refers to multi-node/multi-processor machine learning algorithms and systems designed to improve training performance, accuracy, and scale to larger input data sizes. Data parallelism and model parallelism are different ways of distributing the workload across nodes. Data parallelism is when you run the algorithm in every node but feed it with different parts of the partitioned data; model parallelism is when you use the same data for every node but split the model among nodes.

The following are some frameworks for distributed machine learning.

- Apache Hadoop (HDFS)

- Apache Spark (MLlib)

- Apache Mahout

- TensorFlow and PyTorch: Frameworks with APIs for distributed computation by model/data parallelism

Model Inference Options

Based on the inference implementation, different options are available to achieve the throughput and latency requirements.

- REST endpoints

- Traditional batch inference operations

- Distributed stream processing systems, such as Spark Streaming or Apache Flink

Parallelizing ML inferences across all available nodes/accelerated devices can reduce the overall inference time. Other inference optimization methods batch-size adjustment and model pruning.

The following are popular hardware choices optimized for speed, small size, and power efficiency for scaling in deep learning.

- graphics processing units (GPU)

- field-programmable gate arrays (FPGA)

- vision processing units (VPU)

Input Data Pipeline

With distributed machine learning, multiple powerful GPUs/TPUs consume the processed input data. As a result, the input data pipeline can easily become the new performance bottleneck (see Figure 8-5).

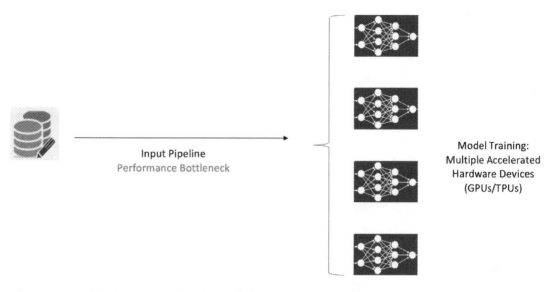

Figure 8-5. *Performance bottleneck: input data pipeline*

A typical input preparation is an ETL pipeline. ETL is short for extract, transform, and load.

- Extract is the process of reading data from multiple and often different types of sources. This stage is I/O intensive.

- Transform is the process of applying rules to convert/combine the extracted data into the desired format. This stage is compute-intensive and typically handled by CPUs.

- Load is the process of writing the data into the target datastore and ready to be fed to the training algorithm. This stage is also I/O intensive.

As shown in Figure 8-6, two instances of the input pipeline are executed sequentially. Each instance feeds a data partition to a GPU/TPU device. There are two gaps: one gap between *transform* operations and another between *train* operations, indicating the CPU and GPU/TPU are idle.

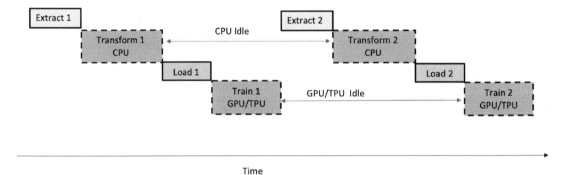

Figure 8-6. *Sequential input data pipeline*

Parts of the input pipeline can be parallelized to make sure the CPUs and GPUs/ TPUs are fully utilized (see Figure 8-7).

- Parallelize file/database reading during the extract phase

- Parallelize transformation operations (shuffle, map, etc.) across multiple CPU cores

- Prefetch pipelines and operations for training

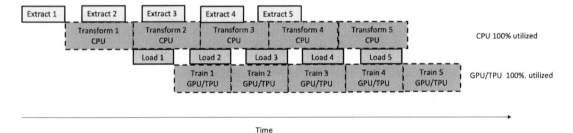

Figure 8-7. *Parallel (optimized) input data pipeline*

ML Tooling Ecosystem

In the beginning of the chapter, we introduced the concept of machine learning pipeline and pipeline automation. As described in Figure 8-2, the machine learning pipeline consists of two distinct phases: Model Development (steps 1, 2, and 3) and Model Deployment (steps 4, 5, and 6).

Figure 8-8 provides an overview of the tooling landscape in the machine learning space, where we group tools according to phases: The Model Development / Training (bottom left) and Model Serving (bottom right) are specific frameworks/libraries/ modules supporting Model Development and Model Deployment phases, respectively.

On top of that, the Machine Learning Platforms (top center) are integrated and centralized services environments. Key functionality of ML platforms includes speeding up data exploration and preparation, accelerating model development and training, and essentially simplifying the scale of ML operations across the entire lifecycle—from model development, experiment tracking, to deployment.

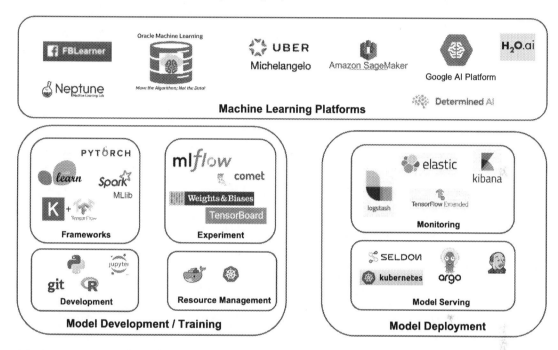

Figure 8-8. *Machine learning tools*

ML Platforms

Below is a list of popular ML platforms in the industry. We will cover more details on the mainstream ML platforms in chapter 9. In chapter 10, we will go over an example that demonstrates ML pipeline automation using Oracle Machine Learning (OML).

- Amazon SageMaker
- Facebook FBLearner
- Oracle Machine Learning
- Uber Michelangelo
- Google AI Platform
- Neptune

Next, let's go over ML Development tools. Tools in the Model Development/Training group can be divided into categories such as ML Frameworks, Development Environments, Experiment Tracking, and Resource Management.

ML Development Tools

Tools in the Frameworks category contains libraries used for modeling and training (see list of examples below). Up until 2011, the ML space was dominated by such libraries.

- TensorFlow

- Keras

- scikit-learn

- PyTorch

- Spark MLlib

In the Development Environments category, Jupyter notebook, Python, R are popular among data scientists. Git is a distributed version control system.

For experiment management, we have mlflow for tracking. Tensorboard is very useful for tracking and visualizing metrics as well as model graphs. More information about tensorboard can be found here: `https://www.tensorflow.org/tensorboard/get_started`.

For resource management, Docker and Kubernetes can be used to define and customize a set of resource usage conditions and policies during data processing and model training.

Since 2016, as large amounts of data and off-the-shelf models become more accessible, the demand for tools to help productize/deploy machine learning increases. As shown in Figure 8-8, the Deployment tools can be divided into Model Monitoring and Model Serving categories.

ML Deployment Tools

Once the trained model is deployed in the production environment, Model Serving tools (such as the following) are useful for simplifying model maintenance, reducing latency, and scale for big data streaming pipelines, etc.

- Seldom

- Argo

- Kubernetes

- Jenkins

- In addition, monitoring tools such as the following are commonly deployed to handle/visualize model drift and data anomaly detection.

- TensorFlow Data Validation

- Logstash

- Elasticsearch

- Kibana

Summary

This chapter analyzed the challenges of deploying and machine learning model-driven applications. We went through key requirements, important concepts, high-level architecture, and design considerations. We discussed machine learning deployment infrastructure and solutions. The chapter ended with a summary of tooling supporting the ML pipeline life cycle.

CHAPTER 9

ML Deployment Pipeline Using Oracle Machine Learning

As large-scale model-driven applications are being deployed at an ever-increasing pace, enterprises and industries are racing to adopt technology that makes the process of building and maintaining these applications more efficient and less expensive. A machine learning (ML) platform is the data center software and hardware infrastructure that automates and accelerates the delivery of the life cycle of predictive applications capable of processing massive data sets. The current trend is to operate the ML platform as a centralized service for the entire enterprise, enabling efficient, uniform, reliable, and reproducible pipelines across many organizations.

This chapter begins with a brief introduction of mainstream ML platforms backed by major AI pioneers (such as Google, Facebook, Oracle, and Uber). We focus on Oracle Machine Learning (OML) as the ideal ML platform for Oracle database professionals to manage machine learning models in production. We dive into OML's built-in high-performance analytical and built-in integration capabilities. This chapter aims to empower database professionals to create and deploy analytics capabilities, scale them to meet expanding business requirements, leverage open source data science packages to manage the complete end-to-end ML life cycle.

Mainstream ML Platforms

ML Platforms centralizes the service and organization of all ML-related tasks. ML platforms can be based on open source or proprietary components. They can be hosted in the cloud or on-premise. However, their core is to run machine learning

© Heli Helskyaho, Jean Yu, Kai Yu 2021
H. Helskyaho et al., *Machine Learning for Oracle Database Professionals*,
https://doi.org/10.1007/978-1-4842-7032-5_9

pipelines. Let's go over some of the major ML platforms and understand the ML pipeline components covered, their dependent frameworks, how they scale, and their target users.

- **Google's TensorFlow Extended (target users: data scientists)**: TFX is an open source TensorFlow-based production-scale Machine Learning platform. TFX is implemented in Google internally, and it intends to support the full ML pipeline. However, it requires the TensorFlow framework. All parallelism and distribution ML strategies target large-scale deep learning model training only. The platform manages the following.

  - Data ingestion

  - Data analysis

  - Data transformation

  - Data validation

  - Trainer

  - Model evaluation and validation

  - Serving

  - Logging

- **Facebook's FBLearner (target users: data scientists)**: FBLearner is an internal ML-as-a-service data center infrastructure that runs on specific FB hardware. FBLearner requires a specific ML framework (such as PyTorch) for model building and ONNX format to exchange models.

- **Uber's Michelangelo (target users: data scientists)**: Michelangelo is built on top of open source components such as HDFS, Spark, Cassandra, MLlib, Kafka, and TensorFlow. It's an internal ML-as-a-service platform that addresses the following stages of the ML pipeline.

  - Manage data

  - Train models

  - Evaluate models

- Deploy models

- Make predictions

- Monitor predictions

- **MLFlow (target users: data scientists)**: An open source machine learning platform from Databricks, a creator of Spark. MLFlow focuses on managing, tracking, packaging, and sharing model training experiments. It requires additional external workflow and tools to handle model serving and scaling in production.

- **Oracle Machine Learning (target users: database professionals, data scientists)**: OML is Oracle's machine learning platform that moves advanced analytics over 30 machine learning algorithms in data management infrastructures such as Oracle Database and Spark 2 Clusters. Along with built-in data security, backup, and restore, the embedded ML algorithms are tuned to deliver scalable, parallel, and distributed executions for model training and model scoring.

The following are the ML pipeline stages directly addressed by OML.

- Data extraction

- Data preparation

- Feature engineering

- Model training

- Model scoring

- Model management/repository

The main goal of this chapter is to provide in-depth coverage of these ML pipeline stages in the context of the Oracle Machine Learning platform. Besides, we explain how multiple OML user interfaces and APIs enable integration with open source packages to address additional stages of the ML pipeline.

- Data validation

- Model monitoring

- Pipeline orchestration and deployment

Oracle Machine Learning Environment

Machine Learning pipeline components are orchestrated using scripting languages such as R and Python. With OML4R and OML4Py (coming soon), pipeline components have immediate access to ML algorithms, data, and model storage in Oracle Database. At the same time, OML4R and OML4Py APIs empower data scientists with open source environments. Figure 9-1 provides an overview of the OML architecture accessible from an ML pipeline.

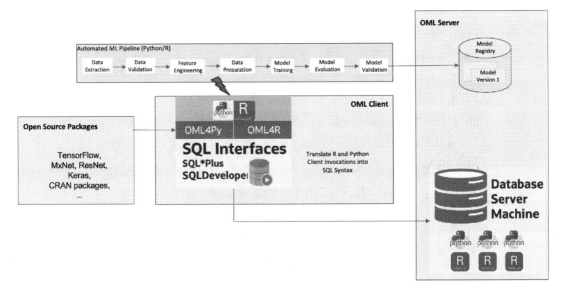

Figure 9-1. *Accessing OML from an automated ML pipeline*

Now let's dive into each ML pipeline component to understand how the OML platform enables them.

Data Extraction in Big Data Environment

Oracle Machine Learning for Spark (OML4Spark) is the R language API component connecting Hadoop to the remaining components of OML, including Oracle Database. As shown in Figure 9-2, OML4Spark acts as a transparency layer running from an R client, OML4Spark uses proxy objects to reference and manipulate (via the standard R syntax) data from various data sources such as file system, HDFS, Hive, Impala, Spark DataFrames, and JDBC sources.

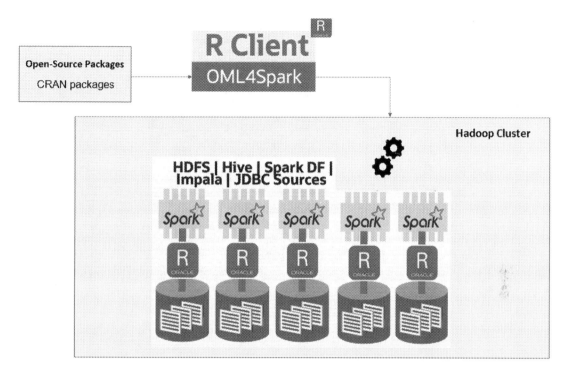

Figure 9-2. *OML4Spark: transparent layer between ML pipeline and big data environments*

In-Cluster Parallel Data Processing

OML4Spark enables scalable data ingestion and preprocessing via a set of R packages and Java libraries. From an R client, these packages/libraries enable direct access and manipulation of the large scale of data stored in a variety of data sources: data stored in a local file system, HDFS, HIVE, Spark DataFrames, Impala, Oracle Database, and other JDBC data sources. Data processing is scalable since it is run across a cluster of Hadoop nodes in preparation in a parallel and distributed manner.

Oracle Cloud SQL enables queries against non-relational data stored in multiple big data sources.

- Apache Hive

- HDFS

- Oracle NoSQL Database

- Apache Kafka

- Apache HBase

- Object Stores (Oracle Object Store and S3)

By combining OML4Spark and Oracle Cloud SQL, you can address large-scale data-driven analytics and discover patterns in big data, relational data, or a combination of both. OML4Spark provides scalable data ingestion and data preprocessing in a distributed cluster as a component of a complex machine learning pipeline.

Automated Data Preparation and Feature Engineering

The ML pipeline typically involves many repetitive activities, such as data preparation, feature engineering, text processing, ensembles, algorithm selection, and model hyperparameter tuning.

General Data Processing Automation

OML Automated Data Preparation (ADP) supports common data transformation tasks such as binning, normalization, and outlier/missing value imputation. ADP can be configured to automatically perform the transformations required by a specific algorithm (more details in Table 9-1). The user can supplement the automatic transformations with additional transformations, or the user can choose to handle all the transformations themself.

Table 9-1. *Transformation Performed by ADP*

Algorithm	Transformation by ADP
GLM	Numerical attributes are normalized.
k-Means	Numerical attributes are normalized.
MDL	All attributes are binned with supervised binning.
Naïve Bayes	All attributes are binned with supervised binning.
NMF	Numerical attributes are normalized.
O-cluster	Numerical attributes are binned with a specialized form of equi-width binning, which automatically computes the number of bins per attribute. Numerical columns with all nulls or a single value are removed.
SVD	Numerical attributes are normalized.

Text Processing Automation

For text processing, OML automatically processes text columns to extract tokens based on TF–IDF scores. The resulting sparse tokenized data is automatically integrated with other features in the provided training data set and provided to the algorithm.

AutoML

AutoML is a new feature being introduced with the coming soon Oracle Machine Learning for Python (OML4Py). It consists of automatic model selection, feature selection, and hyperparameter tuning.

Automated Model Selection

Auto Model Selection identifies the in-database algorithm that achieves the highest model quality faster than an exhaustive search.

Automated Feature Selection

Auto Feature Selection reduces the number of features by identifying the most relevant ones, thus improving model performance and accuracy.

Automated Hyperparameter Tuning

Auto-Tune Hyperparameters significantly improves model accuracy guided by ML and avoids manual or exhaustive search techniques.

Scalable In-Database Model Training and Scoring

In-database model training and scoring algorithms are the most important Oracle Machine Learning features. These algorithms are executed inside Oracle Database. Therefore, they can take full advantage of Oracle Database scalability, parallelism, security, integration, cloud, and structured/unstructured data processing capabilities.

This section examines scale-related capabilities such as in-database parallel execution and in-cluster parallel execution.

In-Database Parallel Execution via Embedded Algorithms

The OML algorithms embedded in the Oracle Database kernel eliminate data movement out of the database.

In this section, let's examine some of the Oracle Machine Learning for R embedded R execution functions that support in-database parallel execution.

As shown in Figure 9-3, the parallel execution capability allows any user-defined script in the ML pipeline to be offloaded to the OML server to take advantage of high-performance computing hardware such as Oracle Exadata Database Machine.

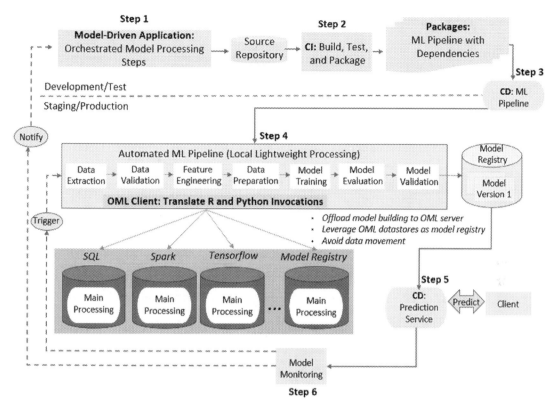

Figure 9-3. *OML-based automated pipeline*

To support embedded R execution, Oracle Database's responsibility can be summarized as follows.

- Managing and controlling the execution of R scripts by spawning potentially multiple server-side R engines at the database server.

- Automatically partitioning and passing data to R engines executing in parallel.

- Eliminating moving data from Oracle Database by referencing data via ore.frame proxy objects.

- Ensuring that all the R engine executions for all the data/task partitions complete; or returning an error.

- Providing the result from the execution of each user-defined embedded R function in an ore.list format. This list is stored in the database until the user requires the result.

The following subsections describe each of these five areas of responsibility in more detail.

Task-Parallel Execution

During a task-parallel execution via the ore.indexApply function, the user-defined function is divided into tasks. The tasks are allocated to one or more R engines for processing. Task-parallel capability can be valuable in various use cases, such as simulations, experimentations, or when an entire model is too large to fit in a single R engine.

Data-Parallel Execution

During a data-parallel execution via the ore.groupApply and rq.groupEval functions, one or more R engines perform the same user-defined R function, or task, on different data partitions. Group-based data-parallel execution enables large numbers of parallel model training sessions, such as hundreds of thousands of predictive models, one model per data segment.

In data-parallel execution for the ore.rowApply and rq.rowEval functions, one or more R engines perform the same user-defined R function on batches (in terms of rows) of data. Row-based data-parallel execution is suitable for scalable model scoring on large data sets.

Degree of Parallelism

The ore.groupApply, ore.rowApply, and ore.indexApply functions rely on their parallel argument to specify the degree of parallelism to use in the embedded R execution. The parallel argument defaults to the ore.parallel option global per ORE connection or FALSE if ore.parallel is not set.

For the SQL equivalent functions, rq.groupEval, and rq.rowEval, the degree of parallelism is specified by the PARALLEL hint in the input cursor argument.

Environments

The rq.groupEval and rq.rowEval functions enable parallel execution in a SQL environment such as SQL*Plus.

The ore.groupApply, ore.rowApply, and ore.indexApply functions enable parallel execution in an R environment.

> **Note**   User-defined R functions invoked using ore.doEval or ore.tableApply are never executed in parallel.

In-Database Parallel Execution with Partitioned Models

As a type of ensemble model, the partitioned models feature in OML automatically partitions data based on one or more columns. OML4SQL builds, organizes, maintains, and persists submodels for the partitions. Submodels are represented as partitions in the top-level model entity, enabling users to easily build and manage many models tailored to independent slices of data. Allowing the user to directly interact with the top-level model also greatly simplifies scoring the partitioned model.

During partitioned model scoring, you must include the partition columns as part of the USING clause. The partition names, columns, and the structure of the partitioned model are stored in the ALL_MINING_MODEL_PARTITIONS view in Oracle Database.

Partitioned models can take full advantage of the in-database parallel execution for model training and scoring of a potentially large number of submodels.

Figure 9-4 describes the process of building a partitioned model.

1. The user specifies a partition key (a.k.a., a comma-separated list of columns) from the input data set.

2. The input data is sliced into multiple data partitions based on the distinct values of the partition key.

3. Each data slice results in its model partition. However, submodels are not standalone models, so they are not visible to the user. The maximum number of partitions is configurable, and the default is 1000 partitions.

4. Submodels are trained using in-database parallel execution.

5. The user directly accesses the top-level model for model scoring.

> **Note**   The partition key only supports column data types of NUMBER and VARCHAR2.

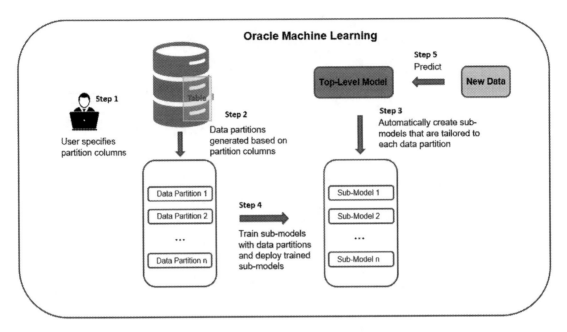

Figure 9-4. Partitioned models: maximizing in-database parallel execution

In-Cluster Parallel Execution

As shown in Figure 9-2, OML4Spark can access the entire Hadoop cluster to perform parallel, distributed machine learning algorithms. These algorithms include the following.

- OML4Spark algorithms such as deep neural, custom linear model (LM), ELM, generalized linear model (GLM), H-ELM, PCA, and SCD. These OML4Spark algorithms are optimized for Spark parallel execution. Therefore, they perform better than Spark MLlib.

- Spark MLlib algorithms such as GLM, LM, gradient boosted trees, PCA, k-means, SVM, LASSO, ridge regression, decision trees, and random forests.

- General parallel, distributed MapReduce jobs run from R, leveraging open source CRAN packages.

Note   Similar core functionality is available in a Java library that can be invoked from any Java or Scala platform.

Model Management

Oracle Machine Learning enables users to build and deploy predictive models inside Oracle Database. To support model management and model analysis, trained models and related information (training/test data, version, parameters, etc.) must be persisted.

If a model is created in Oracle Database, the model and different information related to the model are stored in the following specific database tables and views. The user can query these tables and views to understand the model.

- ALL_MINING_MODELS: This view contains all records for OML models accessible to the current user. Each record has details about the model owner, name, creation date, size, and model algorithm (classification, regression, clustering, etc.), and so on

- ALL_MINING_MODEL_ATTRIBUTES: This view contains information about the columns in the training data used to build the model.

- ALL_MINING_MODEL_SETTINGS: This view describes the parameters used to train the model.

- ALL_MINING_MODEL_VIEWS: This view describes all the model views accessible to the user.

Saving Models Using R Datastores in Database

Since ML pipelines are orchestrated and driven by scripts (such as R) running in a session, you can take advantage of R allows any R objects within a session to be persisted to disk and reloaded across R sessions (see Figure 9-5).

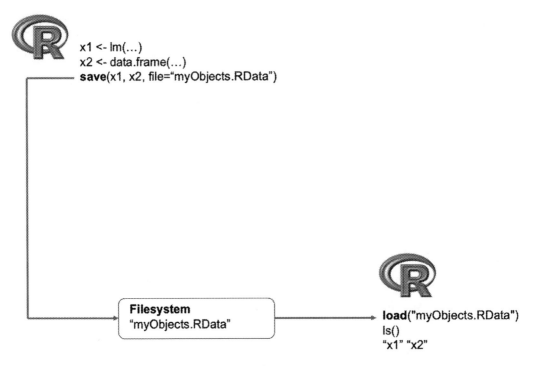

Figure 9-5. *Persisting ML models to the file system*

On the other hand, OML supports object persistence using an R datastore in Oracle Database (see Figure 9-6).

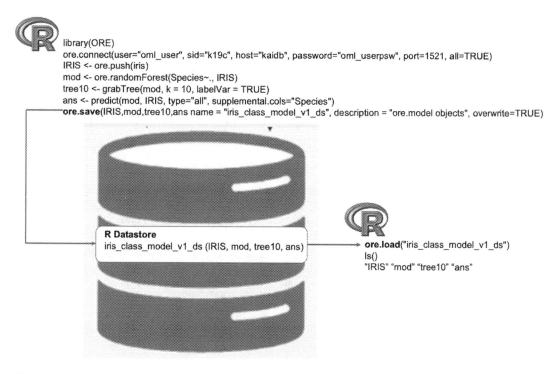

```
library(ORE)
ore.connect(user="oml_user", sid="k19c", host="kaidb", password="oml_userpsw", port=1521, all=TRUE)
IRIS <- ore.push(iris)
mod <- ore.randomForest(Species~., IRIS)
tree10 <- grabTree(mod, k = 10, labelVar = TRUE)
ans <- predict(mod, IRIS, type="all", supplemental.cols="Species")
ore.save(IRIS,mod,tree10,ans name = "iris_class_model_v1_ds", description = "ore.model objects", overwrite=TRUE)
```

R Datastore
iris_class_model_v1_ds (IRIS, mod, tree10, ans)

```
ore.load("iris_class_model_v1_ds")
ls()
"IRIS" "mod" "tree10" "ans"
```

Figure 9-6. *Persisting ML models to R datastore in the database*

This section focuses on how a model registry is implemented using datastores. A model registry keeps track of multiple versions of trained models and the corresponding training/test data sets, features, parameters, and any other metadata required for model scoring, model monitoring, and model management purposes. As a key element of the Oracle Machine Learning platform, Oracle R Enterprise (ORE) supports persistence for trained models (e.g., "mod") along with metadata as R objects onto the database. ORE proxy objects (e.g., "IRIS"), as well as any R objects (e.g., "tree10" and "ans"), can be saved and restored across R sessions as a named entity (e.g., "iris_class_model_v1_ds") in Oracle Database. The model version can be embedded as part of the datastore name. The user has the option to overwrite or delete a datastore.

In Oracle Database, any saved trained model—with all the related metadata, training/test sets, metrics, and hyperparameters—can be retrieved using the datastore name. This persistent mechanism allows any experimentation to be reliably repeated without having to specifically manage files on disk.

Leveraging Open Source Packages

Today, many open source tools are available to address specific demands in the ML life cycle. With the support of multiple APIs such as OML4R, OML4Py, OML4SQL, and OML4Spark, it is possible to incorporate third-party or open source packages with the OML platform to augment ML pipeline functionality. The following sections explain how to enable data validation and model monitoring using open source tools. We cover TensorFlow Extended, scikit-multiflow, and Kubeflow.

TensorFlow Extended (TFX) for Data Validation

TensorFlow Data Validation (TFDV) is a component of TensorFlow Extended, but TFDV can be deployed without requiring TFX.

TFDV validation can be applied to all important features/labels in the input data and predicted values.

Supported anomaly detection use cases are discussed in the following subsections.

Schema-Based Validation

Validity checks are performed according to data types, ranges, and other statistics against a schema. All anomalies are reported as a result of the data validation.

Instead of manually constructing the schema from scratch, the user can automatically use the infer_schema() method to automatically generate the initial version of the schema from any input data set. The user can modify the generated schema as needed.

Training and Serving Skew Detection

This function detects the following types of skew between training and serving data.

- **Schema skew**: Detects if the training and serving data sets do not conform to the same schema.

- **Feature skew**: Detects if the feature values during training are different from feature values at serving time.

- **Distribution skew**: Detects if the distribution of feature values for training data is significantly different from serving data.

Drift Detection

Detection of data drift involves comparing a series of data. Only categorical features are compared between two consecutive data sets (e.g., between consecutive days of training data). The drift is measured in terms of L-infinity distance. The user defines an acceptable threshold distance value so that the user is notified when the drift is higher than the threshold.

scikit-multiflow for Model Monitoring

Based on an algorithm called *adaptive windowing* (ADWIN), the scikit-multiflow Python package can detect concept drift changes in dynamic environments. Whenever a new value is added to the data, ADWIN keeps track of a few statistical properties of data within an adaptive window that grows and shrinks automatically. Statistical properties include the total sum of all values, the window width, and the total variance. The admin. detected_change() method is used to indicate a change. More information can be found in the GitHub repository at `https://github.com/scikit-multiflow/scikit-multiflow`.

Once a change is detected (step 6 in Figure 9-3), the automatic ML pipeline is triggered to restart to train the model with a more recent data set (step 4 in Figure 9-3). Also, more significant drift can cause a notification to be sent to the data scientist if a manual investigation or code change is required. In this case, new code can trigger the redeployment of the entire ML pipeline (steps 1 through 3 in Figure 9-3).

Kubeflow: Cloud-Native ML Pipeline Deployment

Kubeflow is an open source project that enables cloud-native ML pipeline deployment. Kubeflow works in any Kubernetes-conformant cluster (in cloud or on-premise). With Kubeflow, you can do the following.

- Use Jupyter Notebook to develop models.

- Use Kale to convert a Jupyter Notebook to an ML pipeline. The Kubeflow controller automatically spins up and coordinates pods for pipeline steps (see Figure 9-7).

- Use the Kubeflow dashboard to view the pipeline run graph (see Figure 9-8). All pipeline runs are saved for review/rerun (see Figure 9-9).

245

- Use Katib to submit or monitor hyperparameter tuning jobs (see Figure 9-10).

- Use KFServing to create and deploy a server for inference.

- Use custom resources (e.g., TFJob or PyTorch) to simplify distributed training with deep learning frameworks like TensorFlow and PyTorch. Once a TFJob or PyTorch resource is defined, the custom controller takes care of spinning up and managing all the individual pods and configuring them to communicate with one another.

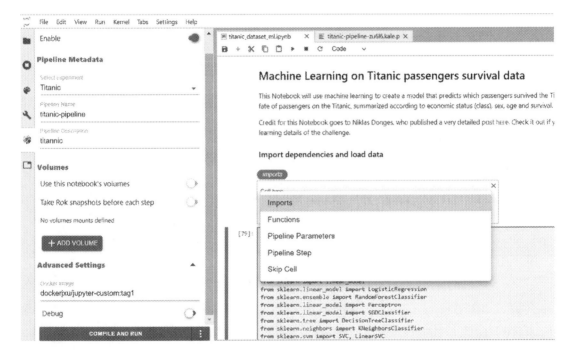

Figure 9-7. *Converting Notebook to ML pipeline using Kale*

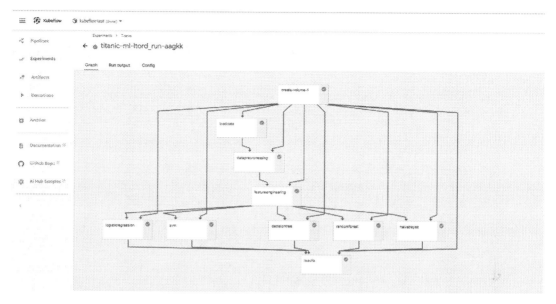

Figure 9-8. *ML pipeline run graph*

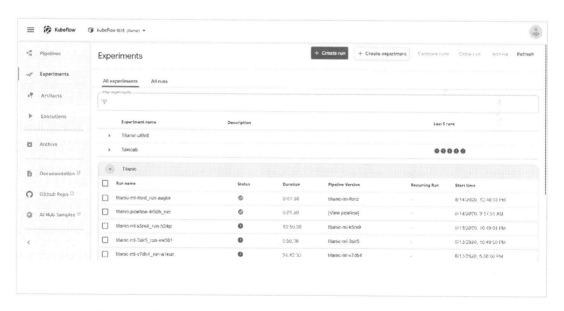

Figure 9-9. *Saved ML pipeline experiments and pipeline runs*

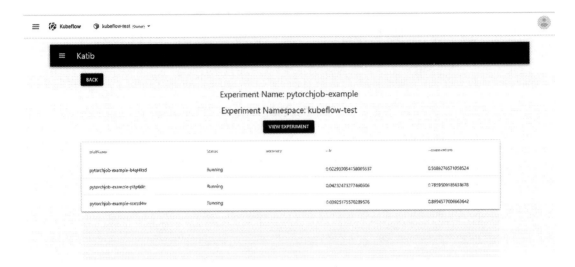

Figure 9-10. *Monitoring hyperparameter jobs with Katib*

Summary

This chapter provided an overview of major machine learning platforms. It explained how Oracle Machine Learning components could support ML pipeline stages.

- Data extraction

- Data preparation

- Feature engineering

- Model training

- Model scoring

- Model management/repository

We introduced multiple OML APIs, such as OML4R, OML4Py, OML4SQL, and OML4Spark. We explained how to use these APIs to incorporate third-party or open source packages (such as TFDV) with the OML platform to augment ML pipeline functionality. We demonstrated how OML's built-in high-performance analytical and built-in integration capabilities enable key requirements of ML productization pipeline: scalability, automation, and integration with open source for end-to-end life-cycle management.

CHAPTER 10

Building Reproducible ML Pipelines Using Oracle Machine Learning

As explained in the previous chapter, Oracle Machine Learning (OML) provides built-in data security and tuned ML algorithms to deliver scalable, parallel, and distributed executions for model training and model scoring. With OML4R and OML4Py, open source components can be integrated into the OML platform to enable additional custom capabilities in the ML pipeline.

This chapter uses an actual supervised machine learning model as an example. It demonstrates how open source packages are integrated into the OML platform to automate the ML pipeline end-to-end. It also demonstrates how to track the versions of all artifacts (source code, data, model, metrics, etc.) of the ML pipeline and make the pipeline experiments reproducible.

As shown in Figure 10-1, the ML pipeline (highlighted in green at the top center) consists of multiple steps: data extraction, data validation, feature engineering, data preparation, model training, model evaluation, model validation, and model monitoring.

In the example covered in this chapter, Oracle Machine Learning (OML) components provide most of the functionality. For example, OML supports data security, scalable model training, and scalable model scoring. In Figure 10-1, these OML components are highlighted in red, and they are deployed as parts of the OML Server (highlighted in gray on the right) or the OML Client (highlighted in gray at the bottom).

For additional pipeline functionality or specific customizations, we incorporate open source packages (highlighted in gray on the left) into the ML pipeline via OML4Py or OML4R (highlighted in red inside the OML Client). In our example, open source

H. Helskyaho et al., *Machine Learning for Oracle Database Professionals*,
https://doi.org/10.1007/978-1-4842-7032-5_10

packages happen to be written in Python, but the OML integration mechanism chosen is OML4R. The Python subprocess module connects to OML client code in Rscript.

The following is a list of the open source packages included in our example.

- TensorFlow Data Validation (TFDV) is used for data validation and model monitoring.

- Data Version Control (DVC) is used for ML pipeline experiment tracking.

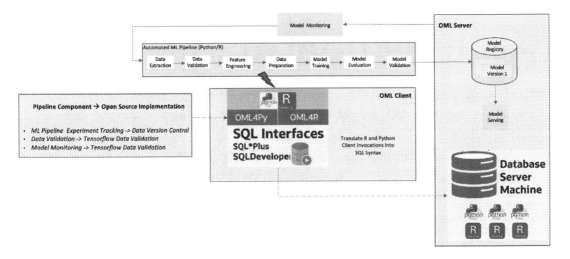

Figure 10-1. *The pipeline implementation: integrating open source components with OML*

TFDV and DVC are discussed more later in the chapter.

The Environment

The main goal of this chapter is to present an ML pipeline example implementing the design as described in Figure 10-1. We cover the software stack of OML client and server and open source components, such as DVC and TFDV. We also go over the main installation and configuration steps. Let's get started.

Setting up Oracle Machine Learning for R

Both the OML4R client and server are set up with Oracle Linux 7 as the operating system. Oracle Database Release 19c is installed on the OML server. Oracle R Enterprise (ORE) 1.5.1 for Oracle Database components are installed on both client and server sides. ORE 1.5.1 is certified with R-3.3.0—both open source R and Oracle R Distribution. R-3.3.0 is compatible with the community-contributed R supporting packages.

The following are the specific R packages and versions to install.

- arules 1.5-0

- cairo 1.5-9

- DBI 0.6-1

- png 0.1-7

- ROracle 1.3-1

- statmod 1.4.29

- randomForest 4.6-12

Note   All the components are included as part of Oracle Database Release 19c for Linux x86-64. You can download it at www.oracle.com/database/technologies/oracle19c-linux-downloads.html/. The versions of these components were the latest at the time of writing this book. You should check for more current versions.

Figures 10-2 and 10-3 show the installation steps on the server side and the client side, respectively.

**Installation Sequence on the
Server Computer**

Figure 10-2. *OML4R server installation sequence*

**Installation Sequence on
the Client Computer**

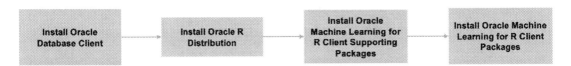

Figure 10-3. *OML4R client installation sequence*

Verifying the Oracle Machine Learning for R Installation

It is always a good practice to verify the software stack for proper configuration. Please follow the following steps to verify ORE library can be loaded and database objects can be accessed from the server and the client side.

Verifying OML4R on the Server Side

First, start an R instance using the ORE script and load the ORE library.

```
% ORE
> library(ORE)
```

A series of "Loading required package: ORExxx" statements should follow without any error.

Next, connect to the server as the OML_USER. The following example connects user oml_user to the database k19c on the kaidb server host using port 1521.

```
ore.connect(user="oml_user", sid="k19c", host="kaidb",
password="oml_userpsw", port=1521, all=TRUE)
```

Now, execute some basic functions.

```
## Is the OML4R client connected to the OML4R server?
ore.is.connected()
[1] TRUE
```

If the connection is successful, the function returns TRUE.

```
## List the available database tables.
ore.ls()
[1] "TABLE1" "TABLE2"
```

If the configuration is successful, the ore.ls() function returns the list of tables in the current schema.

Verifying OML4R on the Client Side

Since our client runs on Linux, start R from the command prompt and type the following command to start OML4R.

```
> library(ORE)
```

A series of "Loading required package: ORExxx" statements should follow without any error.

Use the following command to connect to the OML4R server.

```
> ore.connect(user="OML_USER", sid="pdb1", host="kaidb",
password="OML_USERpsw", port=1521, all=TRUE)
Loading required package: ROracle
Loading required package: DBI
```

Now check connectivity.

```
> ore.is.connected()
[1] TRUE
```

If the connection is successful, the function returns TRUE.

Next, list the ore.frame proxy objects.

```
> ore.ls()
[1] "TABLE1" "TABLE2"
```

If the configuration is successful, the ore.ls() function returns the list of tables in the current schema.

In case of configuration issues, some troubleshooting tips are available in the last section of this chapter.

Once the OML client and server are configured, let's move on to setting up the open source components for the ML pipeline.

Setting up Open Source Components

Our ML pipeline example consists of the OML client and server and open source components. Here, we focus on installing and configuring open source components such as DVC and TFDV.

To install TFDV, run the following command.

```
pip install tensorflow-data-validation
```

To install DVC, run the following command.

```
pip install dvc
```

Alternatively, if you are interested in using the same versions of DVC and TFDV code in our environment, you can skip the previous two `pip install` commands and run the `pip install -r` command with the requirements.txt file attached.

```
pip install -r requirements.txt
```

The following are the contents of requirements.txt file.

```
absl-py==0.10.0
alembic==1.4.2
apache-beam==2.24.0
appdirs==1.4.4
appnope==0.1.0
astunparse==1.6.3
atpublic==2.0
attrs==19.3.0
avro-python3==1.9.2.1
azure-core==1.6.0
azure-storage-blob==12.3.2
```

```
backcall==0.1.0
bleach==3.1.5
bravado==10.6.2
bravado-core==5.17.0
cachetools==3.1.1
certifi==2020.4.5.1
cffi==1.14.0
chardet==3.0.4
click==7.1.2
cloudpickle==1.4.1
colorama==0.4.3
commonmark==0.9.1
configobj==5.0.6
configparser==4.0.2
crcmod==1.7
cryptography==2.9.2
databricks-cli==0.11.0
decorator==4.4.2
defusedxml==0.6.0
dictdiffer==0.8.1
dill==0.3.1.1
distro==1.5.0
docker==4.2.1
docopt==0.6.2
dpath==2.0.1
ds-ml-pipeline==1.0.0
dvc==1.7.3
entrypoints==0.3
fastavro==0.23.6
fasteners==0.15
filelock==3.0.12
Flask==1.1.2
flatten-dict==0.3.0
flufl.lock==3.2
funcy==1.14
future==0.18.2
```

```
gast==0.3.3
gdown==3.12.2
gitdb==4.0.5
GitPython==3.1.3
google-api-core==1.22.2
google-api-python-client==1.12.1
google-apitools==0.5.31
google-auth==1.21.2
google-auth-httplib2==0.0.4
google-auth-oauthlib==0.4.1
google-cloud-bigquery==1.28.0
google-cloud-bigtable==1.5.0
google-cloud-core==1.4.1
google-cloud-datastore==1.15.1
google-cloud-dlp==1.0.0
google-cloud-language==1.3.0
google-cloud-pubsub==1.7.0
google-cloud-spanner==1.19.0
google-cloud-videointelligence==1.15.0
google-cloud-vision==1.0.0
google-crc32c==1.0.0
google-pasta==0.2.0
google-resumable-media==1.0.0
googleapis-common-protos==1.52.0
gorilla==0.3.0
grandalf==0.6
grpc-google-iam-v1==0.12.3
grpcio==1.32.0
grpcio-gcp==0.2.2
gunicorn==20.0.4
h5py==2.10.0
hdfs==2.5.8
httplib2==0.17.4
idna==2.9
ipykernel==5.3.0
ipython==7.15.0
```

```
ipython-genutils==0.2.0
isodate==0.6.0
itsdangerous==1.1.0
jedi==0.17.0
Jinja2==2.11.2
joblib==0.14.1
json5==0.9.5
jsonpath-ng==1.5.2
jsonpointer==2.0
jsonref==0.2
jsonschema==3.2.0
jupyter-client==6.1.3
jupyter-core==4.6.3
jupyterlab==2.1.4
jupyterlab-server==1.1.5
Keras-Preprocessing==1.1.2
Mako==1.1.3
Markdown==3.2.2
MarkupSafe==1.1.1
mistune==0.8.4
mlflow==1.9.1
mock==2.0.0
monotonic==1.5
msgpack==1.0.0
msgpack-python==0.5.6
msrest==0.6.16
nanotime==0.5.2
nbconvert==5.6.1
nbformat==5.0.6
neptune-client==0.4.117
networkx==2.4
notebook==6.0.3
numpy==1.18.5
oauth2client==3.0.0
oauthlib==3.1.0
opt-einsum==3.3.0
```

```
packaging==20.4
pandas==1.1.2
pandocfilters==1.4.2
parso==0.7.0
pathlib2==2.3.5
pathspec==0.8.0
pbr==5.5.0
pexpect==4.8.0
pickleshare==0.7.5
Pillow==7.1.2
ply==3.11
prometheus-client==0.8.0
prometheus-flask-exporter==0.14.1
prompt-toolkit==3.0.5
protobuf==3.12.2
ptyprocess==0.6.0
py3nvml==0.2.6
pyarrow==0.17.1
pyasn1==0.4.8
pyasn1-modules==0.2.8
pycparser==2.20
pydot==1.4.1
Pygments==2.6.1
pygtrie==2.3.2
PyJWT==1.7.1
pymongo==3.11.0
pyOpenSSL==19.1.0
pyparsing==2.4.7
pyrsistent==0.16.0
PySocks==1.7.1
python-dateutil==2.8.1
python-editor==1.0.4
pytz==2019.3
PyYAML==5.3.1
pyzmq==19.0.1
querystring-parser==1.2.4
```

```
requests==2.24.0
requests-oauthlib==1.3.0
rfc3987==1.3.8
rich==7.0.0
rsa==4.6
ruamel.yaml==0.16.12
ruamel.yaml.clib==0.2.2
scikit-learn==0.23.2
scipy==1.4.1
Send2Trash==1.5.0
shortuuid==1.0.1
shtab==1.3.1
simplejson==3.15.0
six==1.14.0
sklearn==0.0
smmap==3.0.4
SQLAlchemy==1.3.13
sqlparse==0.3.1
strict-rfc3339==0.7
swagger-spec-validator==2.7.3
tabulate==0.8.7
tensorboard==2.3.0
tensorboard-plugin-wit==1.7.0
tensorflow==2.3.0
tensorflow-data-validation==0.24.0
tensorflow-estimator==2.3.0
tensorflow-metadata==0.24.0
tensorflow-serving-api==2.3.0
tensorflow-transform==0.24.0
termcolor==1.1.0
terminado==0.8.3
testpath==0.4.4
tfx-bsl==0.24.0
threadpoolctl==2.1.0
toml==0.10.1
torch==1.5.0
```

```
tornado==6.0.4
tqdm==4.49.0
traitlets==4.3.3
typing-extensions==3.7.4.2
uritemplate==3.0.1
urllib3==1.25.9
voluptuous==0.12.0
wcwidth==0.2.3
webcolors==1.11.1
webencodings==0.5.1
websocket-client==0.57.0
Werkzeug==1.0.1
wrapt==1.12.1
xmltodict==0.12.0
zc.lockfile==2.0
```

The Data

As shown in Figure 10-1, any ML pipeline begins with data extraction. An iris flower species data set is the input for our ML pipeline. Iris is a multivariate data set introduced by British statistician and biologist Ronald Fisher in his classic 1963 paper. This data set is available from the OML database. You can also download it from the University of California School of Information and Computer Science Machine Learning Repository at http://archive.ics.uci.edu/ml.

The data set includes three iris species with 50 samples each. The data includes columns with additional properties about each flower. In our example, we build a classification model to predict the species of flowers.

Now that we have set up the open source components (TFDV and DVC) and downloaded the input data (iris), we are ready to demonstrate how to implement ML pipeline functionalities like data validation, model monitoring, and experiment tracking (see Figure 10-1).

Data Validation and Model Monitoring Implementation

Our ML pipeline incorporates open source packages to enable custom functionality. This section covers Data Validation and Model Monitoring implementations based on the TensorFlow Data Validation (TFDV) package.

As shown in Figure 10-1, the automated ML pipeline is fed with an input data set. As the data set is processed and passed through each step, more intermediate data, feature sets, predictions, metrics, and scoring are produced and stored in the Model Registry (inside the OML server in Figure 10-1). TFDV supports data validation by detecting schema or distribution skew in the input data. Similarly, model monitoring is implemented by examining consecutive spans of historical features and predictions using TFDV.

TensorFlow Data Validation (TFDV)

After setting up open source components with `pip install`, the tensorflow_data_validation module is available for importing. As shown in the following Python code, TFDV can generate schema and statistics based on any Pandas dataframe. Anomalies can be detected by comparing a new dataframe against the schema and statistics inferred from an existing dataframe.

```
import numpy as np
import pandas as pd
import os
import tensorflow_data_validation as tfdv
from config import Config

Config.ANOMALIES_PATH.mkdir(parents=True, exist_ok=True)
Config.ORIGINAL_DATASET_FILE_PATH.parent.mkdir(parents=True, exist_ok=True)
Config.DATASET_PATH.mkdir(parents=True, exist_ok=True)

def generateSchema(input_file, schema_file):
    assets=pd.read_csv(input_file,low_memory=False)
    assets_stats=tfdv.generate_statistics_from_dataframe(assets)
```

```
    assets_schema=tfdv.infer_schema(assets_stats)
    tfdv.write_schema_text(assets_schema, schema_file)
    tfdv.display_schema(assets_schema)
    return tfdv.load_schema_text('assets_schema')

def getAnomalies(df, expected_df, anomalies_file):
    assets_stats=tfdv.generate_statistics_from_dataframe(df)
    assets_schema=tfdv.infer_schema(assets_stats)

expected_assets_stats=tfdv.generate_statistics_from_dataframe(expected_df)
expected_assets_schema=tfdv.infer_schema(expected_assets_stats)

    anomalies=tfdv.validate_statistics(assets_stats,expected_assets_schema)
    tfdv.write_anomalies_text(anomalies, anomalies_file)
    tfdv.display_anomalies(anomalies)
    return tfdv.load_anomalies_text('anomalies')
```

Data Validation

Data validation occurs immediately after data extraction. The purpose is to ensure that the input data conforms to an expected schema. The following is sample code.

```
iris_df = pd.read_csv(str(Config.ORIGINAL_DATASET_FILE_PATH))
iris_drift_df = pd.read_csv(str(Config.ORIGINAL_DATASET_DRIFT_FILE_PATH))

getAnomalies(iris_drift_df, iris_df, str(Config.ANOMALIES_PATH /
"anomalies_data_drift.json"))
```

Figure 10-4 shows an example of data change in input data. The original iris data set contains three distinct labels: Iris-setosa, Iris-versicolor, and Iris-virginica. To simulate a significant change in the data, we purposely change one label from Iris-setosa to Iris-setosaaaa.

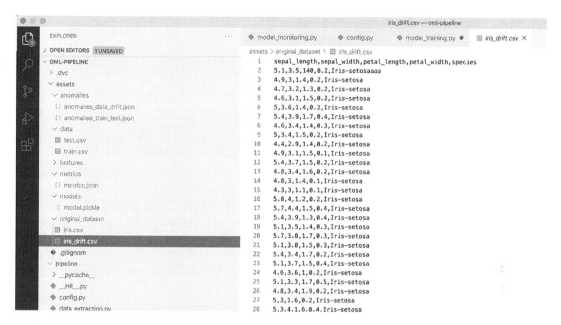

Figure 10-4. *An example of concept drift in the input data*

The output is the anomalies_data_drift.json file. As shown in Figure 10-5, the data validation process first produces a baseline schema based on the original data set. In the schema, the "species" column contains three values: Iris-setosa, Iris-versicolor, and Iris-virginica. Due to the misspelling in the iris_drift_df dataframe, a fourth value, Iris-setosaaaa, is flagged as an error in the anomaly_info block.

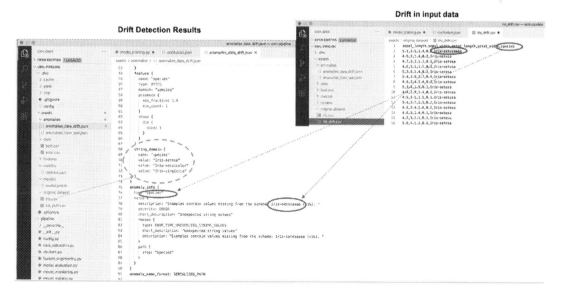

Figure 10-5. *Data validation results: anomalies detected*

Model Monitoring

Model monitoring can be conducted before model serving. The purpose is to ensure the new data is similar to the data used for model training. The following example splits the iris data set into training and test sets, and then compares the two data sets.

```
train_df = pd.read_csv(str(Config.DATASET_PATH / "train.csv"))
test_df = pd.read_csv(str(Config.DATASET_PATH / "test.csv"))

def extract_features(df):
    return df[["sepal_length", "sepal_width", "petal_length",
    "petal_width", "species"]]

train_set = extract_features(train_df)
test_set = extract_features(test_df)

getAnomalies(train_set, test_set, str(Config.ANOMALIES_PATH /
"anomalies_train_test.json"))
```

The resulting anomalies_train_test.json does not contain an anomaly_info block, which indicates no train/test data drift.

Tracking and Reproducing ML Pipeline

TFDV can detect changes in any data/feature/prediction sets. However, not all changes in data have the same impact on the model. Model monitoring must consider changes in areas other than data. For example, changes in the source code and training parameters can require a restart of parts or the entire ML pipeline.

This section demonstrates the implementation of tracking changes and reproducing the ML pipeline using Data Version Control (DVC). With DVC, we can define and customize dependencies among the ML pipeline steps. The dependency relationship drives and triggers the restart of the ML pipeline from the nearest changed step upstream.

Data Version Control (DVC)

DVC is an open source data versioning, workflow, and experiment management software (https://dvc.org). DVC works on top of Git repositories to provide versioning for data and model artifacts and make ML pipeline experiments reproducible.

ML pipeline consists of a series of stages, which typically start with raw data extraction, continue with intermediate feature engineering and model training stages, produce trained models and metrics, and include drift monitoring.

From a command-line interface (dvc run), the user can define an ML pipeline stage. The pipeline definition file (dvc.yaml) can contain the following fields. Figure 10-15 shows a dvc.yaml pipeline definition.

- cmd: The command to be executed in this stage

- wdir: Working directory for the stage command to run in (relative to dvc.yaml file's location)

- deps: Dependency file or directory paths (relative to wdir)

- params: Parameter dependency field names that are read from a file (params.yaml by default)

- outs: Output file or directory paths (relative to wdir)

- metrics: Metrics files

- plots: Plot metrics

- frozen: Whether this stage is frozen from reproduction

- always_changed: Whether this stage is considered as changed by commands such as dvc status and dvc repro. false by default

Versioning Code, Data, and Model Files

The ML pipeline definition in Figure 10-6 and state are stored in a pair of metafiles: dvc. yaml and dvc.lock. These metafiles are small, and they are added to the Git repository along with the source code. Large files and directories such as data sets, features, parameters, metrics, models, and artifacts and stored using the DVC's built-in cache synchronizing with various remote storage.

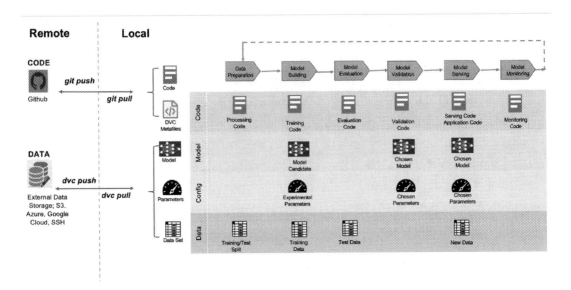

Figure 10-6.   *Code and data flows in DVC*

Demo with Actual ML Pipeline

In our example, the ML pipeline source code entry point is a Python project, oml-pipeline, as shown in Figure 10-7. The pipeline implements data science steps in the following files.

```
pipeline/data_extraction.py
pipeline/feature_engineering.py
pipeline/model_training.py
pipeline/model_evaluation.py
pipeline/model_monitoring.py
ore_client.R
```

Note   In model_training.py and model_evaluation.py, model training and model evaluation are implemented in OML via the ore_client.R script.

As explained in Chapter 9, OML stores the Iris data set, features, metrics, and scoring in the Oracle database. However, to automate the model retrain, we would like to customize dependencies, change tracking, and ultimately, when and how the ML pipeline is restarted.

Let's dive into a simple ML pipeline and demonstrate how easy it is to enable pipeline change tracking and how changes in multiple stages trigger a custom pipeline restart.

ML Pipeline Project with Git and DVC Initialization and Configuration

First, let's change the directory to the ML pipeline project root and run the `git` commands shown in Figure 10-7.

```
(myenv) → oml-pipeline git init
Initialized empty Git repository in /Users/jean/git/oml-pipeline/.git/
(myenv) → oml-pipeline git:(master) ✗ git add .
(myenv) → oml-pipeline git:(master) ✗ git reset
(myenv) → oml-pipeline git:(master) ✗ git add .
(myenv) → oml-pipeline git:(master) ✗ git commit -m "experiment RF"
[master (root-commit) cc485a4] experiment RF
 15 files changed, 614 insertions(+)
 create mode 100644 .gitignore
 create mode 100644 README.md
 create mode 100644 app.py
 create mode 100644 assets/.gitignore
 create mode 100644 ore_client.R
 create mode 100644 pipeline/__init__.py
 create mode 100644 pipeline/config.py
 create mode 100644 pipeline/data_extraction.py
 create mode 100644 pipeline/dbclient.py
 create mode 100644 pipeline/feature_engineering.py
 create mode 100644 pipeline/model_evaluation.py
 create mode 100755 pipeline/model_monitoring.py
 create mode 100644 pipeline/model_training.py
 create mode 100644 pipeline/omlclient.py
 create mode 100644 requirements.txt
```

Figure 10-7. *Git initialization and code check-in*

Next, we run `dvc init` to initialize DVC for the project, `dvc remote add` to specify the default DVC remote storage as a local directory, and `dvc config` to turn off anonymized usage statistics (core.analytics false).

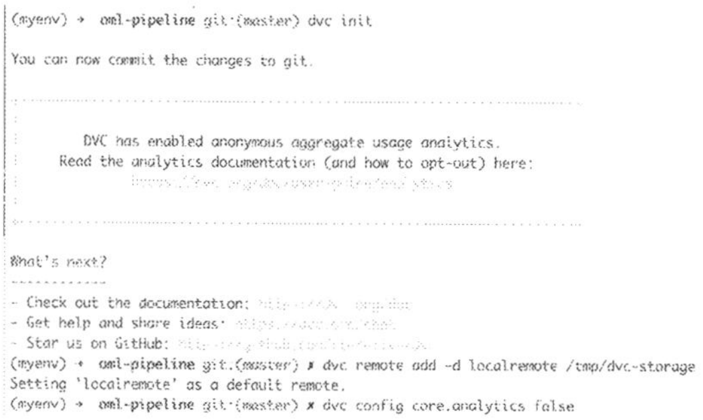

```
(myenv) → oml-pipeline git:(master) dvc init

You can now commit the changes to git.

------------------------------------------------------------

        DVC has enabled anonymous aggregate usage analytics.
     Read the analytics documentation (and how to opt-out) here:
             https://dvc.org/doc/user-guide/analytics

------------------------------------------------------------

What's next?
------------
- Check out the documentation: https://dvc.org/doc
- Get help and share ideas: https://dvc.org/chat
- Star us on GitHub: https://github.com/iterative/dvc
(myenv) → oml-pipeline git:(master) ✗ dvc remote add -d localremote /tmp/dvc-storage
Setting 'localremote' as a default remote.
(myenv) → oml-pipeline git:(master) ✗ dvc config core.analytics false
```

Figure 10-8. *DVC initialization and configuration*

The resulting DVC configuration is automatically saved in the config file in the .dvc directory under the pipeline project.

Figure 10-9. *Autogenerated DVC configuration file*

Defining and Recording Dependencies with DVC

In DVC, a stage is described with a command to execute, data I/O, interdependencies, and results. dvc run is the helper command for creating and updating pipeline stages. Each stage in the DVC pipeline has a dependency on all previous stages. The following are the five stages of the ML pipeline.

1. Extract (data extraction)

2. Feature (feature engineering)

3. Train (model training)

4. Evaluation (model evaluation)

5. Monitor (model monitoring)

Figures 10-10 through 10-14 show the series of dvc run commands that define these five stages.

```
(myenv) → oml-pipeline git:(master) x dvc run -n extract \
-d pipeline/data_extraction.py \
-o assets/data \
"/usr/local/bin/python3 pipeline/data_extraction.py"
Running stage 'extract' with command:
        /usr/local/bin/python3 pipeline/data_extraction.py
Downloading...
From: https://drive.google.com/uc?export=download&id=16Q1UIFxaoZpamSihdsLDuq9p712bvvsz
To: /Users/jean/git/oml-pipeline/assets/original_dataset/iris.csv
100%|████████████████████████████████████████████| 4.62k/4.62k [00:00<00:00, 5.45MB/s]
Creating 'dvc.yaml'
Adding stage 'extract' in 'dvc.yaml'
Generating lock file 'dvc.lock'
Updating lock file 'dvc.lock'

To track the changes with git, run:

        git add dvc.lock dvc.yaml
```

Figure 10-10. *Defining data extraction stage with dvc run*

```
(myenv) → oml-pipeline git:(master) x dvc run -n feature \
> -d pipeline/feature_engineering.py \
> -d assets/data \
> -o assets/features \
> "/usr/local/bin/python3 pipeline/feature_engineering.py"
Running stage 'feature' with command:
        /usr/local/bin/python3 pipeline/feature_engineering.py
Adding stage 'feature' in 'dvc.yaml'
Updating lock file 'dvc.lock'

To track the changes with git, run:

        git add dvc.lock dvc.yaml
```

Figure 10-11. *Defining feature engineering stage with dvc run*

```
(myenv) → oml-pipeline git:(master) x dvc run -n train \
> -d pipeline/model_training.py \
> -d assets/features \
> -o assets/models \
> "/usr/local/bin/python3 pipeline/model_training.py"
Running stage 'train' with command:
        /usr/local/bin/python3 pipeline/model_training.py
before
done
Adding stage 'train' in 'dvc.yaml'
Updating lock file 'dvc.lock'

To track the changes with git, run:

        git add dvc.lock dvc.yaml
```

Figure 10-12. *Defining model training stage with dvc run*

```
(myenv) → oml-pipeline git:(master) x dvc run --force -n evaluation \
-d pipeline/model_evaluation.py \
-d assets/features \
-d assets/models \
-M metrics/metrics.json \
"/usr/local/bin/python3 pipeline/model_evaluation.py"
Running stage 'evaluation' with command:
        /usr/local/bin/python3 pipeline/model_evaluation.py
Modifying stage 'evaluation' in 'dvc.yaml'
Updating lock file 'dvc.lock'

To track the changes with git, run:

        git add dvc.yaml dvc.lock
```

Figure 10-13. *Defining model evaluation stage with dvc run*

```
(myenv) → oml-pipeline git:(master) ✗ dvc run -n monitor \
-d pipeline/model_monitoring.py \
-d assets/original_dataset \
-d assets/data \
-o assets/anomalies \
"/usr/local/bin/python3 pipeline/model_monitoring.py"
Running stage 'monitor' with command:
        /usr/local/bin/python3 pipeline/model_monitoring.py
2020-09-24 18:15:24.565846: I tensorflow/core/platform/cpu_feature_guard.cc:142] This TensorFlow binary is optimized with oneAPI Deep Neural Network Library (oneDNN)to use the
  following CPU instructions in performance-critical operations:  AVX2 FMA
To enable them in other operations, rebuild TensorFlow with the appropriate compiler flags.
2020-09-24 18:15:24.601670: I tensorflow/compiler/xla/service/service.cc:168] XLA service 0x7fb6e01c9150 initialized for platform Host (this does not guarantee that XLA will b
e used). Devices:
2020-09-24 18:15:24.601698: I tensorflow/compiler/xla/service/service.cc:176]    StreamExecutor device (0): Host, Default Version
<IPython.core.display.HTML object>
/Library/Frameworks/Python.framework/Versions/3.8/lib/python3.8/site-packages/tensorflow_data_validation/utils/display_util.py:173: FutureWarning: Passing a negative integer i
s deprecated in version 1.0 and will not be supported in future version. Instead, use None to not limit the column width.
  pd.set_option('max_colwidth', -1)
              Anomaly short description                                          Anomaly long description
Feature name
'species'     Unexpected string values  Examples contain values missing from the schema: Iris-setosaaaa (<1%).
Adding stage 'monitor' in 'dvc.yaml'
Updating lock file 'dvc.lock'

To track the changes with git, run:

    git add dvc.lock dvc.yaml
```

Figure 10-14. *Defining model monitoring stage with dvc run*

The initial execution of the dvc run command line automatically generates a pair of stage tracking files in the working directory: dvc.yaml and dvc.lock. To track the stage changes with Git, run the following.

```
git add dvc.yaml dvc.lock
```

After each subsequent call of dvc run or dvc run, dvc.yaml and dvc.lock are updated to record the changes in the pipeline. Figure 10-15 and Figure 10-16 show the current status of the pipeline with dvc.yaml and dvc.lock files, respectively. This status reflects the pipeline after the series of dvc run commands that define these five stages from Figures 10-10 through 10-14.

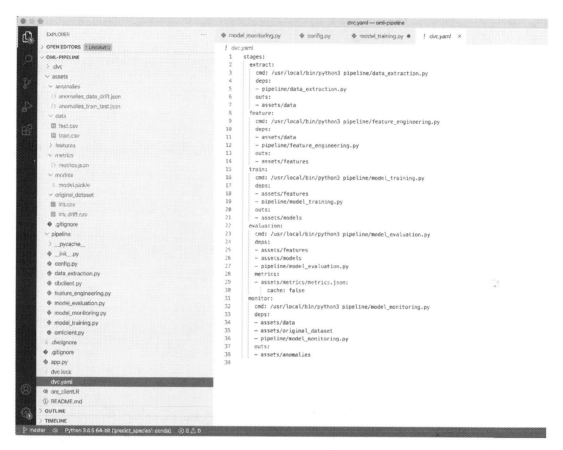

Figure 10-15. *ML pipeline definition (dvc.yaml)*

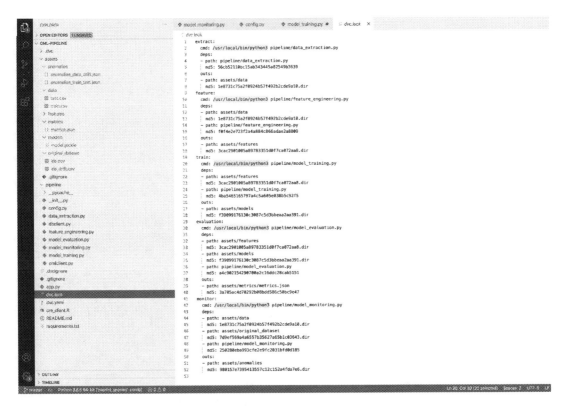

Figure 10-16. *Auto-generated and auto-updated ML pipeline state (dvc.lock)*

Tracking and Reproducing ML Pipelines with DVC

As shown in Figure 10-17, once the generated dvc.yaml and dvc.lock files are added to
the Git repository, changes to all elements of the pipeline stages (I/O, interdependencies,
files, directories, data sets, models, metrics, etc.) are tracked by DVC. From this point on,
DVC can reproduce any version of the pipeline.

```
(myenv) → oml-pipeline git:(master) ✗ git add .
(myenv) → oml-pipeline git:(master) ✗ git commit -m "Complete ml pipeline"
[master b452c19] Complete ml pipeline
 11 files changed, 233 insertions(+), 4 deletions(-)
 create mode 100644 .dvc/.gitignore
 create mode 100644 .dvc/config
 create mode 100644 .dvc/plots/confusion.json
 create mode 100644 .dvc/plots/default.json
 create mode 100644 .dvc/plots/scatter.json
 create mode 100644 .dvc/plots/smooth.json
 create mode 100644 .dvcignore
 create mode 100644 dvc.lock
 create mode 100644 dvc.yaml
(myenv) → oml-pipeline git:(master) git tag -a "rf-pipeline" -m "pipeline with RandomForestRegressor"
```

Figure 10-17. *Track and tag the current state of the pipeline with Git*

Figure 10-18 shows the current metrics.

```
(myenv) + oml-pipeline git.(master) dvc metrics show -T
workspace:
        assets/metrics/metrics.json.
                r_squared: 0.9982459518341872
                rmse: 0.0350126373422287566
```

Figure 10-18. *Display the current model metrics*

The git commit command in Figure 10-17 tracks the current status of the pipeline stages. The git tag -a command adds an rf-pipeline tag that references a specific point in the pipeline change history. As shown in Figure 10-18, the dvc metrics show -T command displays model evaluation metrics.

Now that we tag the pipeline, we essentially establish a "baseline" for tracking future changes.

Sample Tracking Use Cases

Typical use cases of tracking ML pipelines include model drifts caused by changes in the input data (see Figure 10-4) and switching algorithms during pipeline experiments (see Figure 10-19).

Changes in the data set (see Figure 10-4) are detected by DVC because we specified this file as a dependency of two pipeline stages: feature engineering and model monitoring.

Figure 10-19 shows another example of pipeline change tracked by DVC. In the model training code, we can switch the classification algorithms between random forest regression and linear regression. Because the model training code (model_training.py) is part of the training stage definition, a change in the algorithm is flagged by the DVC during pipeline reproduction.

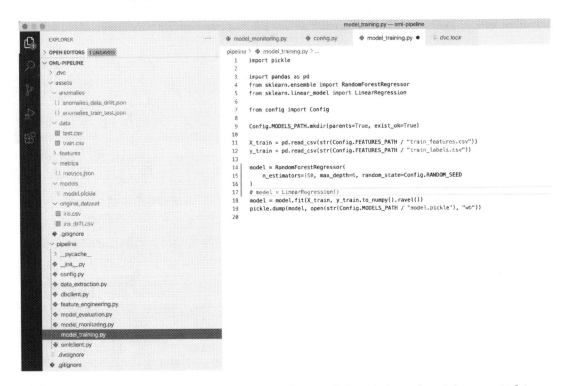

Figure 10-19. *An example of changes in the model training algorithm: switching between random forest regression and linear regression*

Reproducing ML Pipeline: An Example

This section demonstrates how DVC optimizes the pipeline reproduction using the model algorithm tracking information.

Step 1: Update the Pipeline and Track the Change Using Git

In this example, the only change to the pipeline is the classification algorithm switch from Random Forest Regression to Linear Regression. After you update model_training. py with this change, save it.

Step 2: Reproduce the Pipeline Starting from the Evaluation Stage

The dvc repro <stage> command allows the user to specify the stage of the pipeline to reproduce. In this case, after updating the training algorithm, we want to see metrics (produced by model evaluation stage).

In Figure 10-20, DVC is requested to reproduce the ML pipeline starting from the model evaluation stage. With tracking information, DVC discovered that the model training stage has changed. Based on the pipeline definition interdependencies, the data extraction and feature engineering stages are not affected. Therefore, they can be skipped. DVC then automatically runs model training and model evaluation stages. Figure 10-20 also shows the resulting metrics after the reproduction.

This example demonstrates that we can automate all aspects of reconciling ML pipelines regardless of the type/location of the change.

```
(myenv) → oml-pipeline git:(master) ✗ dvc repro evaluation
Stage 'extract' didn't change, skipping
Stage 'feature' didn't change, skipping
Running stage 'train' with command:
        /usr/local/bin/python3 pipeline/model_training.py
Updating lock file 'dvc.lock'

Running stage 'evaluation' with command:
        /usr/local/bin/python3 pipeline/model_evaluation.py
Updating lock file 'dvc.lock'

To track the changes with git, run:

        git add dvc.lock
(myenv) → oml-pipeline git:(master) ✗ git add .
(myenv) → oml-pipeline git:(master) ✗ git commit -m "Complete lr ml pipeline"
[master 5285c63] Complete lr ml pipeline
 2 files changed, 8 insertions(+), 10 deletions(-)
(myenv) → oml-pipeline git:(master) git tag -a "lr-pipeline" -m "pipeline with LinearRegresstion"
(myenv) → oml-pipeline git:(master) dvc metrics show -T
workspace:
        assets/metrics/metrics.json:
                r_squared: 0.9467245149351708
                rmse: 0.19296021497183052
```

Figure 10-20. *Reproducible ML pipeline*

Separate Storage Locations for Code and Pipeline Artifacts

As shown in Figure 10-6, code and pipeline artifacts are stored separately. Figure 10-21 shows the source code changes tracked in Git. Figure 10-22 shows DVC metafiles stored in Git. Figure 10-23 shows other artifacts saved in the DVC built-in cache.

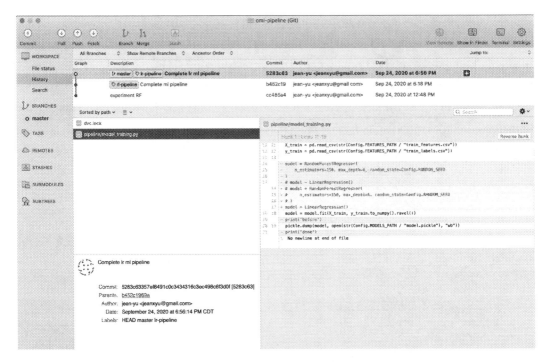

Figure 10-21. *Source code changes tracked as a version of model_training.py in Git*

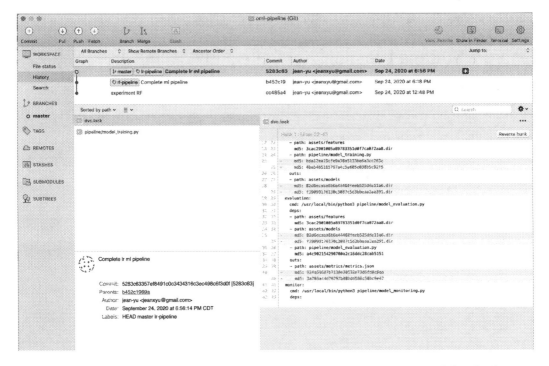

Figure 10-22. *ML pipeline state changes are tracked as a version of dvc.lock in Git*

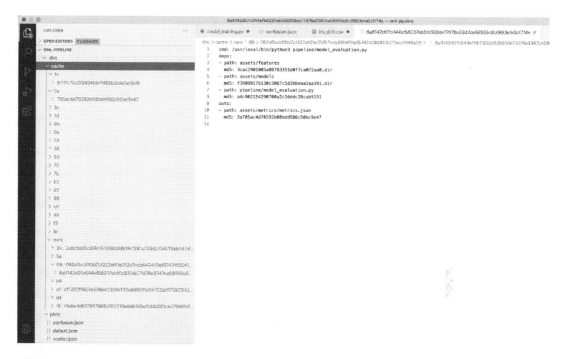

Figure 10-23. *ML pipeline artifacts tracked in DVC built-in cache*

Visualization of ML Pipeline

The dvc dag command can provide a simple visualization of the ML pipeline. Figure 10-24 shows an example of the command's output.

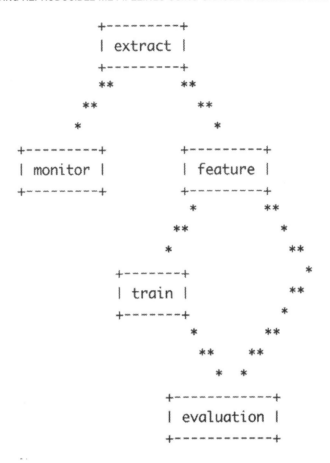

Figure 10-24.   *Visualizing ML pipelines as graph(s) of connected stages*

OML4R Troubleshooting Tips

This section addresses several common issues that can occur during OML4R configuration. Please see the "The Environment" section for additional information.

Error When Connecting to Oracle Database (as oml_user)

Once the OML4R client and server were installed in our demo environment, we loaded the OML4R libraries with the following function call to verify that the basic functionality of OML4R were working.

```
> library(ORE)
```

Next, we made a connection to the OML4R server as OML_USER.

```
ore.connect(user="oml_user", sid="k19c", host="kaidb",
password="oml_userpsw", port=1521, all=TRUE)
```

Initially, we encountered the following error messages.

```
Error in .oci.GetQuery(conn, statement, data = data, prefetch =
prefetch,  :
ORA-28578: protocol error during callback from an external procedure
ORA-06512: at "RQSYS.RQEVALIMPL", line 17
ORA-06512: at "RQSYS.RQEVALIMPL", line 14
ORA-06512: at line 4
```

The error was caused by the fact that the Oracle extproc agent did not allow the required external procedure calls for OML4R. We solved the problem by editing the EXTPROC_DLLS environment variable in the ORACLE_HOME/hs/admin/extproc.ora file.

Solution

Do the following on the server side.

1. Set EXTPROC_DLLS=ANY in $ORACLE_HOME/hs/admin/
   extproc.ora.

2. Restart the database.

Error Due to Missing Packages When Building Models

Once a connection to the OML server was established, we tried simple function calls, such as setting the number of parallel processes as a global option.

```
> options(ore.parallel=8)
```

When we called the ore.randomForest function to load a copy of the training data and build a random forest model, the function call resulted in the following error message.

```
> mod <- ore.randomForest('default.payment.next.month~LIMIT_BAL+AGE+PAY_0+
PAY_2+PAY_3+PAY_4+PAY_5+PAY_6', data = UCI_Credit_Card.train, na.action =
na.exclude)
```

```
Error in .oci.GetQuery(conn, statement, data = data, prefetch =
prefetch,  :
  Error in try({ : ORA-12801: error signaled in parallel query server P001
ORA-20000: RQuery error
Error in .ore.embed.initCairo(pngargs, imglst.name) :
  The package 'Cairo' is required
ORA-06512: at "RQSYS.RQGROUPEVALIMPL", line 121
ORA-06512: at "RQSYS.RQGROUPEVALIMPL", line 118
```

The message seemed to indicate an issue with the required Cairo package. We followed the "Verify Cairo and png Dependencies" section in the Oracle documentation (https://docs.oracle.com/cd/F35732_01/oread/oracle-machine-learning-r-installation-and-administration-guide.pdf) to install Cairo and .png packages, but the error remained the same.

We finally found the problem by directly loading the Cairo library.

```
> library('Cairo')
Error : .onLoad failed in loadNamespace() for 'Cairo', details:
  call: dyn.load(file, DLLpath = DLLpath, ...)
  error: unable to load shared object '/usr/lib64/R/library/Cairo/libs/
Cairo.so':
  libtiff.so.3: cannot open shared object file: No such file or directory
Error: package or namespace load failed for 'Cairo'
```

So the root cause is the required shared library: libtiff.so.3. We resolved the issue by running the following on both OML4R client and server sides:

Solution

```
[root@kaidb downloads]# rpm -i compat-libtiff3-3.9.4-11.el7.x86_64.rpm
[root@kaidb downloads]# rpm -qa | grep libtiff
compat-libtiff3-3.9.4-11.el7.x86_64
libtiff-4.0.3-27.el7_3.x86_64
[root@kaidb downloads]# rpm -ql compat-libtiff3-3.9.4-11.el7.x86_64
```

```
/usr/lib64/libtiff.so.3
/usr/lib64/libtiff.so.3.9.4
/usr/lib64/libtiffxx.so.3
/usr/lib64/libtiffxx.so.3.9.4
```

We also verified other required packages by loading them explicitly using the library(xxx) function.

```
>library(Cairo)
> library(arules)
Loading required package: Matrix

Attaching package: 'arules'

The following objects are masked from 'package:base':

    abbreviate, write

> library(DBI)
> library(png)
> library(randomForest)
randomForest 4.6-12
Type rfNews() to see new features/changes/bug fixes.
> library(ROracle)
> library(statmod)
```

Error When Creating or Dropping R Scripts for Embedded R Execution

We tried to create or drop our own R scripts for embedded R execution to implement more custom functionality. We ran into the following error message.

```
> ore.scriptCreate("Script1")
```

```
Error: role "RQADMIN" is required to perform this operation
```

The RQADMIN role must be explicitly granted to the OML4R user to enable calling ore.doEval() with FUN argument, ore.scriptCreate(), and ore.scriptDrop().

Solution

Log in with system privileges and execute the following statement to grant the RQADMIN role to the OML4R user.

```
SQLPLUS / AS SYSDBA
GRANT RQADMIN TO oml_username;
```

Summary

This chapter demonstrated a scalable and reproducible ML pipeline by integrating open source packages into the OML platform.

S

T

U, V, W, X, Y, Z

Index

A

Accelerated Data Science (ADS), 32

Adaptive windowing (ADWIN), 245

Analytics Cloud (OAC)
- definition, 187
- machine learning, 201, 202
- main navigator menu, 188
- preparation
  - bin/group data, 192, 193
  - business problem, 189
  - connection type selection, 189, 190
  - data and data flow, 191, 192
  - data sets, 193
  - data visualization project, 189
  - home screen creation, 189
  - schedule option, 191
  - sequences, 193
  - transformations/enrichments, 190
- professional edition, 187
- visualization and sharing tools, 188
- visualization/narrate, 194
  - analytics options, 200
  - attribute, 198
  - calculation, 196
  - canvas selection, 199
  - options, 197
  - prediction, 195
  - scenarios/calculation, 194

Analyze and visualize data
- area and line chart, 152
- bar/area/line chart, 151, 153
- bar chart, 150
- customer credit level, 151
- formats, 154
- graphical display icons, 150
- pie chart, 152
- scatter chart, 153, 154
- X axis/Y axis, 150

Artificial intelligence (AI), 1

Auto feature selection, 235

Automatic Data Preparation (ADP), 51, 55, 162

Automation pipeline
- abstraction, 218
- AutoML, 219
- binary classification, 217
- data science steps, 211
- data validation, 217, 218
- implementation, 220
- MLOps, 212–214
- model monitoring, 220
- model registry, 214–217
- scaling solutions, 221
  - accelerators, 221
  - inference implementation, 222
  - input data pipeline, 222–224

© Heli Helskyaho, Jean Yu, Kai Yu 2021
H. Helskyaho et al., *Machine Learning for Oracle Database Professionals*,
https://doi.org/10.1007/978-1-4842-7032-5